Lewis Holmes

The Arctic Whaleman

Winter in the Arctic Ocean

Lewis Holmes

The Arctic Whaleman
Winter in the Arctic Ocean

ISBN/EAN: 9783337253189

Printed in Europe, USA, Canada, Australia, Japan

Cover: Foto ©berggeist007 / pixelio.de

More available books at **www.hansebooks.com**

A WHALE BITING A BOAT IN TWO.

THE ARCTIC WHALEMAN;

OR,

WINTER IN THE ARCTIC OCEAN:

BEING
A NARRATIVE OF THE
WRECK OF THE WHALE SHIP CITIZEN,
OF NEW BEDFORD, IN THE ARCTIC OCEAN, LAT. 68° 10′ N.,
LON. 180° W., SEPT. 25, 1852, COMMANDED BY THOMAS HOWES
NORTON, OF EDGARTOWN, AND THE SUBSEQUENT
SUFFERINGS OF HER OFFICERS AND CREW
DURING NINE MONTHS AMONG
THE NATIVES.

TOGETHER WITH

A BRIEF HISTORY OF WHALING.

BY

REV. LEWIS HOLMES.

BOSTON:
THAYER & ELDRIDGE,
114 & 116 WASHINGTON STREET.
1861.

Entered, according to Act of Congress, in the year 1[illegible]7, by

WENTWORTH & COMPANY,

In the Clerk's Office of the District Court of the District of Massachusetts

STEREOTYPED AT THE
BOSTON STEREOTYPE FOUNDRY.

TO

WHALEMEN,

IN WHOSE EMPLOYMENT, DARING ADVENTURES,

AND MANY DEPRIVATIONS,

THE AUTHOR FEELS A DEEP INTEREST,

This Volume

IS MOST RESPECTFULLY INSCRIBED.

PREFACE.

Of all classes of fishermen, the whaleman takes the precedence. This front position will be readily conceded to him, whether we consider the stupendous object of his pursuit, or the vast extent of waters over which he roams to secure his prey, or the dangers and perils peculiar to his avocation, or the immense pecuniary outlay with which the enterprise is carried on.

Some of the reasons which induced the author to present to the public this narrative containing an account of the wreck of the whale ship *Citizen*, and the subsequent exposure and sufferings of her officers and crew in the Arctic Ocean, are the following: —

1. The instance has never been recorded in the history of marine disaster, in which a ship's company, consisting of *thirty-three* persons, lived so many months among the natives in so high a latitude. 2. Being cast helpless and almost destitute upon such a desolate coast, they had to depend principally upon the kindness and generosity of the natives for protection, food, and clothing.

3. Considering the unfavorable and forbidding circumstances of their condition, in living as the natives lived, and their travels in the depths of winter from one settlement to another in order to avoid starvation, it is remarkable that so many of them, with so little sickness, should be rescued the following year.

A plain statement of these facts the author felt was due to his fellow-townsmen, and would probably be of some considerable interest to all classes of readers, and therefore meriting a permanent record with the varied experience of whalemen.

The limited time the author spent with Captain Norton,* who was then preparing for sea, from whom he received the leading facts in the narrative, after it was concluded to give it to the public, is his only apology for not introducing more extended particulars.

Mr. Abram Osborn, Jr.,† Mr. John P. Fisher,‡ and Mr. John W. Norton,§ now absent at sea, confirmed the report of the captain, besides having contributed important materials to the narrative themselves.

Any information respecting the physical features of the arctic region, and the character of its

* Master of the ship *South Seaman*, of New Bedford.

† Master of the ship *William Wirt*, of New Bedford.

‡ First officer of the ship *General Pike*, of New Bedford.

§ First officer of the ship *William Henry*, of Fairhaven.

inhabitants, is not only deeply interesting, but highly useful. The recent explorations of Dr. Kane, in the American Arctic, have largely increased the bounds of knowledge respecting that remarkable portion of the earth's surface.

Though less attention, perhaps, has been given to the exploration of the Asiatic Arctic, through Behring's Straits, it is, however, a region which is yearly visited by scores of American whalemen, and who have become quite familiar both with its eastern and western coasts, even to the impassable ice barrier, which forbids all further approaches to the north.

The acquaintance which the officers and crew of the *Citizen* formed with the natives during the space of *nine months* in which they lived with them, and thus had so favorable an opportunity to learn their characters and habits, has probably never been surpassed by any other company of men within the present century.

The History of Whaling will give the reader a succinct view of the commencement, progress, and present state of the enterprise. The author would here express his acknowledgments to whalemen who have readily furnished him with many valuable incidents connected with the details of their employment.

L. H.

Edgartown, June, 1857.

ILLUSTRATIONS.

CONTENTS.

CHAPTER I.

CHAPTER II.

CHAPTER III.

CHAPTER IV.

CHAPTER V.

CHAPTER VI.

CHAPTER VII.

CHAPTER VIII.

CHAPTER IX.

CHAPTER X.

CHAPTER XI.

CHAPTER XII.

CHAPTER XIII.

HISTORY AND DETAILS OF WHALING.

CHAPTER I.

CHAPTER II.

CHAPTER III.

CHAPTER IV.

CHAPTER V.

CHAPTER VI.

CHAPTER VII.

INTRODUCTION.

A FATHER once said to his son, respecting books, "Read first the introduction; if that be good, try a few pages of the volume; if they are excellent, then, but not else, read on." But I do not wish this criterion to be applied in the present instance. For if the reader find the introduction uninteresting, he will be compensated by a careful perusal of the narrative itself. It may be relied on as stating matters of fact. The information it contains respecting the adventurous and exciting business of the whale fishery is derived from authentic sources. The volume presents matters of deep and general interest to every reader. It will remind him of some of the scenes so vividly portrayed by the late Dr. Kane in his arctic explorations.

Many "that go down to the sea in ships, and do business in the great waters," come from remote parts of the country. Here is the information which will convey to relatives at home some just idea of the toils and privations of those loved ones who are ploughing the trackless ocean. The young men, who are looking forward to a life on the ocean wave, will read the following narrative with eagerness and delight. Their ardent temperament and roving disposition have pictured in fancy's halls bright scenes on the briny deep. Such will find in this work a true view of a mariner's life, accompanied with valuable counsels.

It is neither, as I judge, the tendency nor the design of the book, to deter any from a seafaring life that love adventure, and believe there is no royal road to fortune. To employ a nautical phrase, "None need expect to creep in by the cabin windows; all must crawl through the hawser hole." He must endure hardship and privation before he can enjoy promotion. Young men of sound health, steady purpose, moral courage, and trustworthiness, will, by the blessing of Providence, be sure of promotion. If, however, these qualities are lacking in a young man, the discovery is generally made during his first voyage. His reputation, good or bad, will reach home long before the ship returns to port. Owners and agents know what is in him, and what may be expected from him in future. If he stands the test, if he is faithful and prompt in the discharge of duty, all who have an interest in the success of voyages will want his services.

This book gives valuable information to parents whose sons are inclined to go to sea. Resistance should not be carried to such a degree as to drive the young man to expedients in order to get away from home clandestinely. This course always throws him into the hands, and places him under the power, of those who have no ultimate object but to make all possible profit out of his toil. Many young men, in consequence of obstacles thrown in their way, never divulge at home their longing desire to try the sea. Hence, some, who are physically unfit to bear the fatigue, have taken their first step by running off to some seaport; and after being involved in expense, although they may have changed their mind, they cannot retrace their steps. Once shipped, they will have one life-long regret. Let parents do all they possibly can to render their would-be sailor sons independent of the "landsharks."

Clergymen and Sabbath school teachers are in a position to know something of the tendencies and aptitudes of their respective charges. They may, by availing themselves of the contents of this volume, prevent many mistakes and unavailing regrets. The object to be sought is, to secure those who will go to sea from doing so under false apprehensions of the kind of life, and the essential qualifications for success in the sailor's avocation.

Seamen are in demand, and if commerce continues to extend, many more able-bodied sailors will be required to man our ships. When the extent and value of the interests involved are thought of, it seems surprising that efforts are not made to improve the character and condition of the sailor. Millions of property are intrusted to his care. Thousands of precious lives are in his hands for weeks and months; yet many sailors are the refuse of jails, penitentiaries, and state prisons. The sentiment too often prevails that the worse man makes the better sailor. Hence we may easily account for many shipwrecks, vessels cast away, sunk, and burned. This is becoming too expensive. As total abstinence on board ship has reduced the rates of insurance, so will greater security to life and property be experienced at sea, when more attention shall be paid to the character and condition of sailors. The profit may be proportionally divided between the owners and the crew.

Combined and earnest efforts should be made by parties, especially by those more immediately concerned. A beginning must be made somewhere. Individual effort has done something, but these attempts are limited by others who have it in their power to hinder the operation of beneficial changes. Difficulties can be overcome effectually when all combine to remove them. The temporal and spiritual welfare of so many thousands of our fellow-men afloat on the great deep demand much at our hands.

The moral and religious element must lie at the foundation of all physical and social reforms. When, therefore, owners, captains, agents, officers, and crews will agree not to lower their boats, even if a fish is seen, on Sabbath, an important step will be taken in the right direction. For while all are systematically violating the law of God, there can be no law on board ship but brute force. Why should any one feel at liberty to prosecute his daily employment at sea, when he would be ashamed to do so on land? Is the sailor less dependent on the blessing of a gracious God than the husbandman? He is a very godless farmer who will plough or sow on Sabbath because it rained on Saturday, or may rain on Monday. The difficulties are not insuperable. Meet them at the outset. The responsibility must not be wholly devolved on the captain, to do what he thinks best after he has gone to sea. Let the owners distinctly and unequivocally express their will in the case. Then captain, officers, and crew will be shipped with this stipulation: No fishing on Sabbath.

Many captains and others now engaged in the whaling fleet will welcome such an arrangement. The effect of it on the whole ship's company will be salutary. As the business is now conducted, there are doubtless many uneasy consciences. Some are glad when no fish is seen on Sabbath. But when the cry is raised, "There she blows!" what a struggle takes place in the mind of the pious and God-fearing men! But the rest think, if the boats are not lowered, that their rights and interests are infringed; and even the owners might afterwards complain that, when fish were seen, they were not taken. So the order is given, "Lower away the boats." But this does not settle the question, for the captain feels his moral power diminished. He cannot next day with a clear conscience read and pray with his officers, nor call all hands together next Sabbath to hear the word of God read.

Thus nearly all that is done for the moral improvement of sailors in port is neutralized by one act of disobedience to God.

In New Bedford, something is attempted for the good of the sailor. "The Sailor's Home" is well conducted. The Port Society sustains the Bethel and its indefatigable minister. All the Bibles needed for the ships come from the New Bedford Bible Society. But something further is required to induce habits of Bible reading on board ship. Let owners and others think of some of the hints given above.

If this volume, now presented to the public, containing a narrative of so much interest, will improve the condition of the weather-beaten sailor, and promote the honor of God, our desire is accomplished.

J. GIRDWOOD.

NEW BEDFORD, MASS., *June*, 1857.

THE POLAR BEAR.

THE WHALEMAN;

OR,

NINE MONTHS IN THE ARCTIC.

CHAPTER I.

Ship Citizen sails from New Bedford. — Captain, Officers, and Crew. — Interest centred in a Whale Ship. — Accompanying Ships. — Seasickness and Homesickness. — Arrival at Cape Verd Islands. — An Agreement with Captain Sands, of the Ship Benjamin Tucker. — Whales raised. — Christmas Supper on board of the Citizen. — A Whale Scene. — "An ugly Customer." — A Whale Incident, copied from the Vineyard Gazette. — Arrival at Hilo. — Sandwich Islands.

The whale ship Citizen, of New Bedford, owned by J. Howland & Co., fitted for three or four years, and bound to the North Pacific on a whaling voyage, sailed from the port of New Bedford, October 29, 1851. She was commanded by Thomas Howes Norton, of Edgartown, Martha's Vineyard.

Her officers were the following, namely: first mate, Lewis H. Roey, of New Bedford; second mate, John P. Fisher, of Edgartown; third mate, Walter Smith, of New Bedford; fourth mate, William Collins, of New Bedford. Four boat steerers, namely: Abram Osborn, Jr., and John W. Norton, of Edgartown, John Blackadore and James W. Wentworth, of New Bedford.

The following were nearly all the names of her crew: Charles T. Heath, William E. Smith, Christopher Simmons, George W. Borth, Darius Aping, William Nye, Manuel Jose, Jose Joahim, Charles C. Dyer, Charles Noyes, Edmund Clifford, George Long, Charles Adams, Bernard Mitchell, Nicholas Powers, William H. May, Alpheus Townshend, Barney R. Kehoe, Joseph E. Mears, James Dougherty, and Peter M. Cox. The whole number on board when she sailed was thirty-three persons. In addition to the above, five seamen were shipped at the Verd Islands, which made thirty-eight, all told.

As is generally the case, the majority of these were strangers, and perhaps had never seen each other's countenances until they appeared on the deck of the ship, henceforth to be their new home for months, and it may be for years.

Besides, in this number there were representatives from different and distant sections of the country, and not unfrequently an assortment of nations, and even races.

Here were gathered for the first time many a wandering youth, attracted to the seaboard by the spirit of romantic adventure, to see the world of waters, and to share in the excitement of new scenes. His wayward history, in breaking away from the wholesome restraints and watchcare of home, may be found written, perhaps, in many sorrowful hearts which he has left behind. Years may pass away before either parents or relatives shall hear again from the absent one, and it may be never. Such instances are not uncommon.

How much interest there is centred in a whale ship, as she is about to leave port! It is felt not only by those who embark their property and lives in her, but there are other attractions towards the ship. They are found in the desolateness which is felt in many home circles, in bidding adieu to husbands, sons, and brothers. When the anchor is weighed, and the sails are spread to the faithful breeze, sadness reigns in many households and in many hearts. The thoughts are not only painfully busy concerning *present* separations, but they bound forward to the future, and anticipate what may be the experience of a few years to come. Changes! one hardly dares think of them! Amid the perils and dangers of the deep, how long will the ship's company remain unbroken? Will the *ship* ever return, and reënter her port again? Will those who have just released themselves from the embraces of friends, and wiped away the falling tear, and barred their hearts to the separation, will *they* ever return? or, if they should, will they ever see again those whom they are now leaving? These inquiries and reflections find expression only in painful emotions, sadness, and sorrow. Time will make changes, and leave its ineffaceable footprints with every passing year.

The land was lost sight of in the evening of the day upon which we sailed, with a strong south-west wind. We were accompanied out of the bay by two other outward bound whale ships — the Columbus, of Fairhaven, Captain Crowell, and the Hunter, of New Bedford, Captain Holt.

After the usual passage, with variable winds, and no particular incident of marked importance, except the ordinary and certain amount of seasickness on board, which generally attends the uninitiated in their first interviews with "old Neptune," Cape Verd Islands were made on the 4th of December.

With seasickness, homesickness follows; and then it is that many of the inexperienced, having left good homes and quiet life, wish a thousand times that they had never "learned the trade." But all such wishes are now in vain. With a new life on shipboard and in the forecastle, romance passes away, and leaves in its place the stern outlines of a living reality. Seasickness, however, is only a temporary affair; in most cases, indeed, it soon subsides, and then spirits and hope revive with recruited and invigorated health.

We took our departure from the islands on the 6th, in company with the ship Benjamin Tucker, Captain Sands; strong breezes, north-east trades. The first whales were seen about lat. 30° S., lon. 31° 41′ W., distant about seven miles — light winds. We set signal for the Benjamin Tucker, four or five miles distant, to notify Captain Sands that whales were in sight — an agreement we made while sailing in company. Boats were lowered; the mate fastened to a whale, which brought the shoal to. The second mate was less successful; his boat was stoven by a whale, and his men were floating about upon scattered and broken pieces of the wreck. Other boats soon came up and rescued their companions. The ship now ran down to the boat which was fastened to the whale. The whale, however, was lost, in consequence of cutting the line in the act of lancing him. After a pursuit of an hour or more, the mate fastened to another whale, and finally secured it, though it proved to be of but little pecuniary value. At the same time the boats of the Benjamin Tucker captured a whale, but they could not boast of much superiority. It made them *three* barrels. Thus ended the first whaling scene on the voyage, and certainly not a very profitable day's work.

The Citizen was put on her course. We passed several ships — weather good. December 20, lat. 40° S., whales were raised again, but took no oil. Still in company with the Benjamin Tucker. On Christmas Eve, Captain Sands and his wife took tea on board of our ship, thus reviving remembrances of home and friends, though thousands of miles distant from our native port.

The next incident of more than ordinary interest was another whale scene, of sufficient excitement and peril to satisfy the most ardent and aspiring.

The Benjamin Tucker had luffed to, headed to the westward, with signal to the Citizen that whales were in sight. The ship Columbus was then in company. The three ships were in full pursuit of the monsters of the deep. The school was overtaken in course of an hour or two working to the leeward. At first, one of the boats was lowered from the Citizen, and then another, and another, until four boats were bounding over the waves, each seeking to be laid alongside of his victim, and join in the uncertain conflict. From the three ships there were twelve boats pressing forward with the utmost celerity to share in the encounter, and each emulous to bear off his prize. The fourth boat despatched from the Citizen fastened to a whale. He was shortly lanced, and spouted blood — a sure indication that he had received his death wound. In mortal agony, he plunged, and floundered, and mingled the warm current of his own life with the foaming waters around him. Conscious, apparently, of the authors of his sufferings, with rage and madness he at once attacked the boat, and with his ponderous jaws seized it, and in a moment bit it in two in the centre. Nor was there any time to be lost by the humble occupants of the boat. The rules of courtesy and ordinary politeness in entertaining a superior were for the time being laid entirely aside. Each seaman fled for his life — some from the stern, and others from the bow, while the cracking boards around and beneath them convinced them that the whale had every thing in his own way. Besides, the sensation was any thing but pleasant in expecting every moment to become fodder to the enraged leviathan of the deep. In quick succession those enormous jaws fell, accompanied with a deep, hollow moan or groan, which evinced intense pain, that sent a chill of terror to the stoutest hearts. They felt the feebleness of man when the monster arose in his fury and strength. A boat was soon sent to the rescue of their companions, who were swimming in every direction, to avoid contact with the enraged whale, which seemed bent on destroying every thing within his reach. He really asserted his original lordship in his own native element, and was determined to drive out all intruders. He therefore attacked the second boat, and would probably have ground it to atoms, had not a fortunate circumstance of two objects perhaps somewhat disconcerting him, and dividing his attention, turned him off from his purpose.

The captain of the Citizen, observing the affray from the beginning, was soon convinced that matters were taking rather a serious direction, and that not only the boats but the lives of his men were greatly imperilled. He therefore ordered the fifth boat to be instantly lowered, manned with "green hands," the command of which he himself assumed, and directed in pursuit of the whale. Five boats were now engaged in the contest, with the exception of the one stoven, and all the available crew and officers, including the captain, concentrated their efforts and energies in order to capture this "ugly customer." Just at the moment he was attacking or had already attacked the second boat, the captain's boat appeared on the ground, and from some cause best known to himself, the whale immediately left the former and assailed the latter. What the whale had already done, and what he appeared determined still to do, were by no means very flattering antecedents, and would very naturally impress the minds of "green hands," especially, that whaling, after all, was a reality, and not an imaginary affair or ordinary pastime.

On, therefore, the whale came to the captain's boat, ploughing the sea before him, jaws extended, with the fell purpose of destroying whatever he might chance to meet. As he approached near, the lance was thrust into his head and held in that position by the captain, and by this means he was kept at bay, while the boat was driven astern nearly half a mile. In this manner he was prevented from coming any nearer to the boat, the boat moving through the water as fast and as long as he pressed his head against the point of the lance. This was the only means of their defence. It was a most fortunate circumstance in a most trying situation. If the handle of the lance had broken, they would have been at the mercy of a desperate antagonist. The countenances of the boys were pallid with fear, and doubtless the very hair upon their heads stood erect. It was a struggle for life. It was death presented to them under one of the most frightful forms. They were, however, as singularly and as suddenly relieved as they were unexpectedly attacked. The whale caught sight of the ship, as was supposed, which was running down towards the boats, and suddenly started for the new and larger object of attack. This was observed by the captain, who immediately made signal to keep the ship off the wind, which would give her more headway, and thus, if possible, escape a concussion which appeared at first sight inevitable. The whale started on his new course towards the ship with the utmost velocity, with the intention of running into her. The consequences no one could predict; more than likely he would have either greatly disabled the ship, or even sunk her, had he struck her midships. To prevent such a catastrophe —

the injury of the ship, and perhaps the ruin of the voyage — every thing now seemed to depend upon the direction of the ship and a favoring wind. Every eye was turned towards the ship; oars were resting over the gunwale of the boats, and each seaman instinctively fixed in his place, while anticipating a new encounter upon a larger scale, the results of which were fearfully problematical. A good and merciful Providence, however, whose traces are easily discernible in the affairs of men both upon the ocean and upon the land, opportunely interfered. The ship was making considerable headway. The whale started on a bee line for the ship, but when he came up with her, in consequence of her increased speed before the wind, he fell short some ten or twelve feet from the stern. The crisis was passed. On he sped his way, dragging half of the boat still attached to the lines connected with the irons that were in his body. His death struggle was long and violent. In about half an hour he went into his "flurry, and turned up." Colors were set for the boats to return to the ship; the dead whale was brought alongside, cut in, boiled out, and seventy-five barrels of sperm oil were stowed away.

We copy the following whale incident from the *Vineyard Gazette* of October 14, 1853. The editor says, —

"We are indebted to Captain Thomas A. Norton, of this town, one of the early commanders of the whale ship Hector, of New Bedford, for the following interesting particulars relative to an attack upon and final capture of an ugly whale. Captain Norton was chief mate of the Hector at the time.

"'In October, 1832, when in lat. 12° S., lon. 80° W., the ship ninety days from port, we raised a whale. The joyful cry was given of "There she blows!" and every thing on board at once assumed an aspect of busy preparation for the capture. The boats were lowered, and chase commenced. When we got within about three ships' lengths of him, he turned and rushed furiously upon us. He struck us at the same moment we fastened to him. He stove the boat badly; but with the assistance of sails which were placed under her bottom, and constant bailing, she was kept above water. The captain, John O. Morse, came to our assistance. I told him he had better keep clear of the whale; but he said he had a very long lance, and wanted to try it upon the rascal. Captain Morse went up to the whale, when all at once he turned upon the boat, which he took in his mouth, and held it "right up on end," out of the water, and shook it all to pieces in a moment. The men were thrown in every direction, and Captain Morse fell from a height of at least thirty feet into the water. Not being satisfied with the total destruction of the boat, he set to work and "chewed up" the boat kegs and lantern kegs, and whatever fragments of the boat he could find floating on the water. At this stage of the "fight," I told Captain Morse that if he would give me the choice of the ship's company, I would try him again. It was desperate work, to all appearance, and up to this time the vicious fellow had had it all his own way. The captain was in favor of trying him from the ship, but finally consented for us to attack him again from a boat. With a picked crew, we again approached the whale, now lying perfectly still, apparently ready for another attack, as the event proved. Seeing our approach, he darted towards us with his mouth wide open, his ponderous jaws coming together every moment with tremendous energy. We gave the word to "stern all," which was obeyed in good earnest. As we passed the ship, I heard the captain exclaim, "There goes another boat!" She did go, to be sure, through the water with all speed, but fortunately not to destruction. The monster chased us in this way for half a mile or more, during most of which time his jaws were within six or eight inches of the head of the boat. Every time he brought them together, the concussion could be heard at the distance of at least a mile. I intended to jump overboard if he caught the boat. I told Mr. Mayhew, the third mate, who held the steering oar, that the whale would turn over soon to spout, and that then would be our time to kill him. After becoming exhausted, he turned over to spout, and at the same instant we stopped the boat, and buried our lances deep in "his life." One tremendous convulsion of his frame followed, and all was still. He never troubled us more. We towed him to the ship, tried him out, and took ninety barrels of sperm oil from him.

"'When we were cutting him in, we found two irons in his body, marked with the name of the ship Barclay, and belonging to the mate's boat. We afterwards learned that three months before, when the same whale was in lat. 5° S., lon. 105° W., he was attacked by the mate of the ship Barclay, who had a desperate struggle with him, in which he lost his life.'

"Captain Norton, at the time of the adventure with this whale, had 'seen some service,' but he freely confesses that he never before nor since (though he has had his buttons bitten off his shirt by a whale) has come in contact with such an ugly customer as 'the rogue whale,' as he was termed in sailor parlance. He seemed to possess the spirit of a demon, and looked as savage as a hungry hyena. Our readers may imagine the effect such an encounter would have upon a crew of 'green hands.' During the frightful chase of the boat by the whale, their faces were of a livid whiteness, and their hair stood erect. On their arrival at the first port, they all took to the mountains, and few, if any of them, have ever been seen since."

The Citizen was put on her course again, with strong breezes and fair wind. About five days after, we spoke with the Benjamin Tucker, but Captain Sands had taken no oil. In lat. 47° S. another whale was raised; three boats were lowered in pursuit, but before he could be reached by the irons, he turned flukes, and was seen no more. Lost sight of the Benjamin Tucker. We shaped our course for Statan Land. In lat. 48° S. we experienced a very heavy gale from the south-west, which continued with great severity for twenty-four hours. We spoke with the bark Oscar, Captain Dexter, bound round the cape.

Statan Land in sight, passed seventeen ships, all bound for the cape. The Citizen was eleven days in doubling the cape, and experienced very heavy weather. In lat. 54° S. we raised the first right whale, but, blowing hard, could not lower.

Whales were in sight several days in succession, but we could not lower, on account of rugged weather. In lat. 47° S. a ship was discovered with her boats down in pursuit of whales; came up with her; lowered for right whales, and chased them for an hour or more, but took none. At this time we spoke with the ship Columbus again, with one of her boats fastened to a whale. She had one boat stoven.

Passed St. Felix Islands, on the coast of Chili, and sighted the Gallipagos. In crossing the equator, it was calm for twenty-seven days, and but little progress was made during that time. On the 20th of April, 1852, after a passage of more than five months from New Bedford, we entered the port of Hilo.

Hilo is a port on the Island of Hawaii, one of the cluster of islands in the North Pacific Ocean called Sandwich Islands. They were discovered by Captains Cook and King in 1778, who gave them their present name, in honor of the first lord of the admiralty. The group consists of ten islands, but all of them are not inhabited; they extend from lat. 18° 50′ to 22° 20′ N., and from lon. 154° 53′ to 160° 15′ W., lying about one third of the distance from the western coast of Mexico to the eastern coast of China. By the census of 1849, the population of seven of the islands is given as follows: Hawaii, 27,204; Oahu, 23,145; Maui, 18,671; Kauhai, 6,941; Molokai, 3,429; Nuhua, 723; Lanai, 523; amounting to 80,641.

Most of these islands are volcanic and mountainous. In several places the volcanoes are in activity. Some of the mountains are of great height, being estimated at fifteen thousand feet.

The climate is warm, but not unhealthy, the winter being marked only by the prevalence of heavy rains between December and March. A meteorological table gives as the greatest heat during the year, 88° of Fahrenheit; as the least, 61°. Some of these islands are distinguished for the cultivation of the yam, which affords quite a valuable supply for ships.

The situation of the Sandwich Islands renders them important to vessels navigating the Northern Pacific, and especially so to whalemen. The ports of Hilo, Lahaina, Honolulu, and a few others, are the resort of a large number of whale ships, for the purpose of obtaining recruits. They may be considered as a central point, where ships meet both in the fall and spring, and from whence all matters of intelligence are transmitted to San Francisco, and from the latter place to the Atlantic States.

Formerly all ship news and letters were brought

from the islands to the Atlantic States by homeward bound ships around the Horn, which required for their passages from three and a half to five months. But now, in consequence of mail communications across the isthmus to San Francisco, and from thence to the islands, letters and other public intelligence from the last-named place reach us in six weeks or two months from date.

CHAPTER II.

Recruited for the Arctic. — Departure. — Coast of Kamtschatka. — Copper Island. — Going into the Ice with Captain Crosby. — Gale of Wind. — Dangerous sailing in the Ice. — Cape Thaddeus. — Bay of the Holy Cross. — Plover Bay. — Dead Whale. — St. Lawrence Bay. — Whales working north. — Loose, floating Ice. — Ice covered with Walrus. — Fine Weather. — Striking an Iceberg. — Ship leaking. — Return to St. Lawrence Bay. — Damage repaired. — Arrival in the Arctic.

At the port of Hilo the ship was recruited for the Arctic. We remained in port fifteen days, sailed for Honolulu, and left letters for owners and for home. We touched at another port before proceeding to the north, and there we took in an additional supply of provisions, and then directed our course towards the straits.

In lon. 180° W. we hauled to the north towards the coast of Kamtschatka. Passed Copper Island. We saw many ships on our passage thus far, but we took no whales until June.

About this time we captured two whales off shore, and found great quantities of ice. Spoke with Captain Crowell, of the ship Columbus, and Captain Crosby, of the ship Cornelius Howland.

We went into the ice with Captain Crosby, in search of whales, and soon found them; boats were lowered; pursuit commenced; several were struck; but our irons drew, and we therefore lost them.

A gale of wind coming on and increasing, we worked out of the ice as soon as possible. We were at that time, when the gale commenced, some fifteen or twenty miles in the floating and broken masses, of varied thickness and dimensions, greatly obstructing the course of the ship, and rendering her situation at times exceedingly dangerous. But by constant tacking and wearing in order to avoid concussion with the ice, or being jammed between opposite pieces, both ships were finally worked out of the ice in safety.

On the inside of Cape Thaddeus, we saw a large number of ships; spoke with several, but they reported that whales were scarce.

We now put the ship on her course for Behring Straits. We took one whale off the Bay of the Holy Cross, which made the fourth since we left port. We sailed along the coast towards the east; land frequently in sight; foggy; heard many guns from ships for their boats.

When off Plover Bay, ten miles from land, we picked up a dead whale, having no irons in him, nor anchored, and therefore a lawful prize. Many dead whales are found by ships in course of the season, and especially when ice is prevalent. They are struck by different boats, and if in the vicinity of ice, they will surely make for it, and go under it or among it; under these circumstances the lines must be cut. After some time, the badly wounded whales die, and are picked up as before stated.

We passed between St. Lawrence Island and the main land, or Indian Point. The huts of the natives were plainly seen from the ship's deck; still working our way towards the straits. At this time, we were in company with the ship Montezuma, Captain Tower, and the ship Almira, Captain Jenks. Whales were seen going towards the north, as it is usual for them to do so at this season of the year.

We anchored in St. Lawrence Bay; weather foggy. The natives came off to trade, and brought their accustomed articles for traffic, such as deer and walrus skins, furs, teeth, &c. They take in exchange needles, fancy articles, tobacco, &c.

After a few days, the fog having cleared away somewhat, we stood towards the north again; heard guns; saw whales; still in company with afore-mentioned ships; blowing heavy; all the ships in sight were under double-reefed topsails; beating.

Passed East Cape. Saw whales, but they were working quickly to the north; we followed them in their track with all the sail we could carry on the ship; they came to loose, floating ice, into which they went and shortly disappeared. A novel, and yet a common sight was now witnessed; the ice was covered with a vast number of walruses, which, to appearance, extended many miles.

The weather being fine for the season, the last part of June, in company with the Almira, Captain Jenks, we concluded we would go into the ice again, and if good fortune would have it so, we might capture a few whales.

Accidents occur not unfrequently when least expected, and sad ones, too, arise sometimes from the slightest circumstance, or inattention. Contact with icebergs, or large masses of block ice, when a ship is under sail, is highly dangerous. A momentary relaxation of vigilance on the part of the mariner may bring the ship's bows on the submerged part of an iceberg, whose sharp, needle-like points, hard as rock, instantly pierce the planks and timbers of a ship, and perhaps open a fatal leak. Many lamentable shipwrecks have doubtless resulted from this cause. In the long,

WALRUS AMONG ICEBERGS.

heavy swell, so common in the open sea, the peril of floating ice is greatly increased, as the huge angular masses are rolled and ground against each other with a force which nothing can resist.

The striking of the Citizen against a mass of ice, which nearly resulted in the loss of the ship and the destruction of the voyage, was simply inattention or misunderstanding the word of command.

The man at the wheel was ordered not to "luff" the ship any more, but "steady," as she was approaching a mass of ice; indeed, ice was all around us, which would have passed us on our larboard bow, and thus we should have escaped a concussion; but instead of doing this, he put the wheel down, which brought the ship into the wind, and the consequence was, a large hole was stoven in her larboard bow; the ship began to leak badly. Casks were immediately filled with water, and placed on the starboard side of the ship, and thus in a measure heeled the ship, which brought the leak to a considerable extent out of the water; otherwise, she must have sunk in a very little time. So far as we were able, we temporarily repaired the injury, and made all possible sail on the ship, in order to seek some place of safety, where the whole extent of the damage could be ascertained.

4

In the present disabled and crippled condition of the ship, we felt it was exceedingly perilous and unsafe to remain even a single day in the Arctic. We therefore left the whale ground, and though our progress was slow, yet we put upon the ship all the sail she would bear, since on account of the leak she was very much heeled, and we were obliged to sail her in that condition.

Nor was it safe for our ship to be left alone to beat her way back two hundred miles or more, unaccompanied by another vessel, lest by some unforeseen circumstance, — an event not altogether improbable, — the ship might founder at sea, and all on board perish.

Captain Jenks, of the ship Almira, therefore, kindly proffered his services, with whatever aid he could give, and accompanied our ship nearly to the point of her destination, to the Bay of St. Lawrence, which was about two hundred miles distant from the place where the accident occurred.

When off East Cape, we obtained some plank from the ship Citizen, Captain Bailey, of Nantucket. We passed the heads of the bay, and, with shortened sail, we worked our way up more than thirty miles beyond the direction of any chart, our boats being sent ahead, and sounding the depth of water. We finally reached a point, and came to anchor in a little basin, or inlet, about one hundred and sixty feet from the shore, in five fathoms of water, completely landlocked.

Here in good earnest we commenced breaking out the fore hold abreast of the leak, and took out casks, shooks, &c., and careened the ship still more, which exposed at once the full extent of the damage which the ship had sustained from the ice.

It was found that several planks and timbers were badly stoven. Repairs were made with the utmost expedition; and in seven days from the time the ship went into the bay, she was out again, and on her way towards the north, as strong, and perhaps stronger than she was before.

We passed through the straits, and came to anchor north of East Cape, in company with the ship E. Frazer, Captain Taber, and the bark Martha, Captain Crocker. After lying there three or four days, we got under weigh and stood towards north by west, with high winds, and foggy. We heard whales blowing in the night. The next day whales were seen going north; we followed, and finally passed the "school." We changed the course of the ship, beat back, found them again, and commenced taking oil.

About the first of August, the fog having cleared away, we saw a large number of ships "cutting in" and "boiling out," actively engaged in securing a good season's work. We took several whales at this time. All were busy, and at work as fast as possible, in capturing whales, cutting and boiling. The whole scene, in which were some forty or fifty ships taking whales and stowing away oil, was one of exciting and cheering interest.

Such times as these are the whalemen's harvests.

On the 15th of August, during a heavy blow, we lost run of the whales. We spoke with several ships about this time, among which were the Benjamin Morgan, Captain Capel, and the General Scott, Captain Alexander Fisher.

From this last date to the 22d of September, we spoke with a great number of ships; sometimes whales were plenty, and at other times scarce; and the weather equally changeable; sometimes heavy blows, rainy, and foggy; and then again mild and pleasant.

Among others we spoke with Captain Henry Jernegan, and Captain John Fisher, both of whom are now no more, having finished their earthly voyages, and gone to their "long home."

CHAPTER III.

Northern Lights. — High Winds. — Spoke with Captain Clough. — Ships seen in the Distance. — Storm increasing. — No Observations. — Blowing heavily. — Scene awfully sublime. — Ship struck by a Sea. — Shoal Water. — Rocks and Breakers. — Ship unmanageable. — Fore and mizzen Topsails carried away. — Ship striking astern, bow, and midships. — Foremast cut away. — Narrow Escape of Captain Norton. — Mizzenmast gone by the Board. — Sad Condition of the Seamen. — Land in Sight. — Ship drifting towards the Shore. — Undertow. — The Lantern Keg. — Mainmast cut away, and falling towards the Shore. — Men escaping on the Mast. — Trying Scene. — Captain washed ashore. — Affecting Deaths. — Wreck piled up on Shore. — Fire made. — Men perishing with Cold — Five missing. — Prospects dark. — Destitution. — Tent erected. — Merciful Circumstances connected with the Wreck.

On the 21st of September, we finished cutting in a whale, about twelve o'clock, midnight, wind high from the north-east. The northern lights were uncommonly brilliant, which prognosticated a storm; and the broken water and flying spray round the vessel seemed as if composed of an infinite number of diamonds glistening in the rays of the sun.

The season of the year had now arrived in which, in those high latitudes, sudden changes and violent storms were expected. At three o'clock on the morning of the 22d, the ship was put under short sail; rough; unable to keep fires in the furnace; ship heading to the south-east. We spoke with Captain Clough, who had just taken in a "raft" of blubber. We took a whale; and for a little time the wind moderated, which gave us hope that we should have favorable weather some time longer. Captain Clough left us that day, and turned his ship towards the straits, saying, "I am bound out of the ocean, and have enough." His ship was full; he had thirty-two hundred barrels of oil on board.

We concluded to remain on the ground a while longer, in lat. 68° N. The wind, which had in a measure subsided, now began to rise and increase, until it had reached a heavy gale. We saw in the distance several ships steering for the straits, and bound for the islands. On the 23d, it blew hard, and we were unable to boil.

We judged we were, at this time, about one hundred and fifty miles from land. The weather had been thick for several days past, and therefore we were unable to get an observation. We saw several ships lying to, and heading some one way and some another. The water, we perceived, was very much colored, which indicated that we were drifting towards the eastern shore of the Arctic. At twelve o'clock, wore ship, heading north-west by north. At the same hour that night, wore ship again, heading north-east.

We passed a ship, within the distance of half a a mile, under bare poles, laboring very hard. On the 24th, four o'clock, wore ship north-north-west, wind blowing very heavily from the north-east. We saw great quantities of drift stuff, such as barrels, wood, &c., probably the deck load of some ship swept by the sea. At twelve o'clock, wore ship again; the wind appeared to lull somewhat, but the sea was very rugged; we judged we were about one hundred or one hundred and twenty miles distant from land; weather thick, with rain, sleet, and fog. About one o'clock, on the morning of the 25th, the wind increased, and swept over the ocean with the violence of a hurricane. The darkness of the night added to the tumultuous and mountainous waves that were running at that time; the surface of the ocean lashed into fury by the thickening storm, still gathering its strength; the noble ship now rising the crested billow, and then sinking into the watery valley beneath, and pressed down by the beating and overwhelming elements, made the scene one of indescribable grandeur and awfulness. With the return of morning light, an ugly sea struck the ship, and took her spars from the bow, and carried away one of the starboard boats.

The mate immediately reported to the captain, who was below at the time, that the ship was in shoal water. As soon as he reached the deck, he ordered to set the fore and mizzen topsails. About the same time, the fourth mate reported that there were rocks and breakers just before and under the bows of the ship. From the house, the captain saw projecting rocks through the opening waters, and land all around to the leeward, while the sea was breaking with tremendous violence between the ship and shore.

It now became a certainty, which no earthly power could change, that the ship must go ashore; and the only hope for any one on board was to avoid, if possible, the fatal reef, which appeared to extend out some distance from the land. To strike upon that reef was certain destruction; we saw no way of escape.

The man at the wheel was ordered to put the helm hard up, and at the same time command was given to the seamen to sheet home the fore topsail. The ship immediately paid off two or three points, when she was struck again by another sea, that threw her round on the other tack.

The ship was now in the midst of the rollers,

pitching and laboring dreadfully, while the sea was flying all over her deck, and the spray reaching nearly or quite to her fore and main yards. She was utterly unmanageable; and, at this instant, another sea boarded her, and took off three boats. The yards were ordered to be braced round as soon as possible; but, in the act of bracing them, a terrible blast of wind struck and carried away the fore and mizzen topsails, half-sheeted home. The foresail was now ordered to be set, the ship still pitching, tumbling, and rolling frightfully, and tossed about as a mere plaything at the mercy of winds and waves. In the act of setting the foresail, the weather clew was carried away, and with the next sea the ship struck aft very heavily, and knocked her rudder off, and sent the wheel up through the house. From five to eight minutes she struck forward with such stunning and overwhelming effect that the try-works started three or four feet from the deck, and opened a hole so large in her starboard bow that the largest casks came out.

About this time, the foremast was cut away, with the hope of temporarily relieving the foundering vessel. Shortly after this, the ship struck midships; and the dreadful crash which followed showed that her entire framework was shattered, while the standing masts bent to and fro like slender reeds when shaken by the wind. This was in effect the finishing blow; and what was to be done towards rescuing any thing below deck must be done soon or never.

The captain, at this critical juncture, went into the cabin to secure what articles he could, such as clothes, nautical instruments, money, &c. While there, the stern burst in, and the water came in between the opening timbers in such torrents as to send him backward and headlong with the few articles he had hastily gathered, and scattered them in every direction. The floor of the cabin opened beneath his feet. There was no time for delay. His life was in imminent peril. He at once started for the deck, but was unable to reach it on account of the house having been thrown down upon the gangway, and the mizzenmast having gone by the board, one part of which rested upon the rail. All access to the deck by the cabin doors was thus cut off.

Mr. Fisher became aware of the condition of the captain in the cabin, and called to him to come to the skylight; and as he jumped, he was caught by his arms, and drawn up by several who had come to his rescue. On reaching the deck, the captain saw at once the sad condition of his men. The sea was making a clear breach over the vessel, and they were huddled together round the forecastle and forward part of the ship, amazed, stupefied, cold, and shivering, and had apparently given themselves up to the fate which awaited them.

The fog having in a measure cleared away, the land was more plainly seen, and just at hand — not more than three hundred yards distant. The mainmast was still standing; and there was every indication that the entire top of the vessel, including the first and second decks, had become separated from her bottom, and was drifting in towards the shore. This proved to be the case. The standing mast was now inclining towards the shore, which seemed to present the only way to deliverance and life. The captain, therefore, encouraged his men to seize the first opportunity which should occur, and escape to land, and the sooner they did so the safer and better.

As the ship changed her position by the action of the waves, which swept over and around her with resistless fury, the end of the flying jib boom, at one time, was brought quite near the shore. The seamen were again urged to make an effort to save themselves. It was, indeed, a most desperate chance to venture an escape even from a present danger, with the liability of falling into another, unknown, and perhaps more to be dreaded. Though so near the solid land, towards which every eye looked and every heart panted, still the surging billows and receding undertow around the bow of the ship, were sufficient to appall the most courageous mind.

About this time, as near as can be recollected, the cooper and one of the boat steerers, having dropped themselves from the bow, reached land in safety. The captain, having observed that two had gained the shore, and knowing the utter impossibility of getting fire ashore if it was deferred until the breaking up of the ship, and without it all must unavoidably perish, even if they were saved from a watery grave, held up the lantern keg to attract their attention, and, making signs to them as far he was able for them to look after and save it, tossed the keg overboard. It was borne on the advancing and retreating waves back and forth for more than a quarter of a mile, before it was finally secured. In this keg, which was air-tight, there were candles, matches, tinder, and other combustible materials. It was indeed a most timely and fortunate rescue.

An effort was now made to get a line ashore. One of the crew fastened a line round his body, and attempted to reach the shore, the captain paying out the warp as was necessary. But in consequence of the great force of the current and undertow around the bow of the ship, the line swayed out so far that the man was compelled to let it go in order to save his life. It was with the greatest difficulty he reached the shore.

As the only and last resort which remained, offering reasonable prospect of deliverance, the mainmast was cut away. The ship was now lying nearly broadside to the shore, with her deck inboard, and so much heeled that it required the greatest attention to prevent one from falling off. The mast fell in the direction of the shore, and nearly reached land. The sea was still breaking with fearful power over the vessel, and its spray flying in dense masses over every thing around us, and the din of the thundering billows, as they beat upon the wreck and upon the shore, drowned all human voices to silence.

Again the captain passed along to the forward part of the ship, and once more remonstrated, urged and entreated his men to exert themselves for their safety and lives, as they had now the same means of getting ashore that the officers had; and, furthermore, that in a short time the deck would go to pieces, and then there would be but little, if any hope of their being saved. He resolved he would not leave the wreck until he saw his men in a fair way of escape. Up to this time, no one, it is supposed, had been lost; several had reached land in safety, but those still on the wreck were exposed every moment to a watery grave.

At length, the steerage boy lowered himself down from the bow, and with manly efforts sought to gain the land. He was immediately swept away, and was never seen after. About this time, many began to crawl down on the mainmast, still lying in the direction of the shore. In working their way along on the mast, their progress was not only slow, but they were chilled, benumbed with cold, their clothes thoroughly wet to their backs, and the sea at the same time flying over them. It was with the greatest difficulty they could hold on. The sight was a most affecting one. It was a period of painful anxiety. How many of these seamen will be saved? — how many will be lost?

While attempting thus to escape upon the mast, the advancing or the returning waves would frequently wash numbers off, and then they would struggle with all their energies to regain the mast or the rigging; while those who were more fortunate, and had retained their hold, would aid them as far as possible in getting on to the mast again. It was a most trying and heart-rending scene.

The captain and Mr. Fisher were on the quarter deck, and observed a part of a boat hanging

WRECK OF THE CITIZEN.

by the side of the ship; and they proposed to get into it, and, if possible, reach the land. Their purpose was to hold on to the boat, and thus be borne by the sea towards the shore. They did get into it; but whether it was carried towards the shore or-not, or what became of the piece of the boat, they have no recollection. They were struck by a sea, and probably stunned. The first returning consciousness the captain had, he found himself floating alongside of the ship. He knew not what had become of Mr. Fisher until some time after. He regained a foothold on the quarter deck again, and seemed awakened more fully than ever to the conviction that he must do something, and that soon, in order to save his own life. He was chilled, benumbed, and exhausted; chances of escape appearing less and less probable, as a last resort, said Captain Norton, "I threw myself into the water, among casks, broken pieces of the wreck, and, besides, my own men floating all around me, that I might, if possible, gain the shore. I was probably insensible for some time. I knew nothing of what took place around me. When I came to myself, I found I was lying near the edge of the water, having been cast ashore by some friendly wave. I looked around, and the first man I saw was the fourth mate, floating about in the water a short distance from me. Mr. Fisher was washed ashore about the same time I was. We hastened to the fourth mate as soon as we were able; and one held on to the hand of the other, and hauled him ashore, supposing him to be dead. He, however, revived."

A heavy sea came along, and washed a number from the mast, and brought them ashore; but one man was carried off by the undertow outside the ship. The next sea brought him near to the shore again; and four of those on shore took hold of each other's hands, and ventured as far as safety would allow into the water, and succeeded in drawing him safe to land.

The condition of the carpenter was painful and distressing in the highest degree; yet no one could help him — no earthly power could afford him any assistance. He was plainly seen by those on shore. He was probably washed from the mast, with some others, and carried out to the deck again; and while there, he was doubtless caught in between the opening planks and timbers, and held fast by his legs; and it may be he was otherwise injured. He answered no signs made to him from the shore; he made no effort to free himself or to escape; and, in his case, an escape was an impossibility. In that position, his head dropped upon his breast, and there he died. Soon after, another sea struck the deck, and broke it all to pieces. The largest part that could be seen was that from the bow to the fore chains.

Another painful occurrence was witnessed by those on the shore. A Portuguese sailor was discovered floating about among the broken pieces of the wreck, among casks, barrels, &c. His efforts for self-preservation were remarkable. His shipmates would most gladly have given him a helping hand, but it was impossible to do so. Every heart was moved with sympathy for him. As the towering wave would hurl towards him some piece of the wreck, or a cask or barrel, he was seen to dive, and thus avoid being crushed by it. This he did repeatedly, until, from exhaustion or injury, or both, he sunk to rise no more.

We had three dogs on board, but they were all either killed or drowned; and of three hogs, only one got ashore alive. Within two hours from the time the ship first struck, the wreck was piled up on shore, opposite to where the disaster occurred, to the height of ten feet or more. Spars, timbers, planks, casks both whole and broken, shooks, &c., were thrown together in frightful confusion; and in this promiscuous mass we saw what was once our home and hope on the deep. Here we saw before our eyes a striking illustration of the feebleness of man's frail bark, and with what ease it is torn to pieces, and scattered far and wide, by the resistless power of the elements.

All who were living of our number had reached the shore. Those that were saved had become greatly chilled, and some were nearly perishing. Notwithstanding it was storming at the time, one of the first efforts of a part of our men was to make a fire over a cliff some little distance from the shore, affording a partial protection from the wind and rain.

In searching for articles as they came ashore, we discovered a small keg of spirits, which, in our condition of cold and destitution, was somewhat reviving to all our minds. Five casks of bread, also, were cast upon the beach; but neither beef nor pork was found. The latter probably sunk where the ship left her bottom.

The whole company was soon gathered round the fire, in order to dry our clothes, and, if possible, to obtain some additional warmth. All, however, of our former number were not there; it was a solemn gathering, and the appearance of all of us indicated that we had a narrow escape. Alas! some of our comrades and fellow-seamen were left behind in the surges of the deep, or mingled with the floating wreck, or cast with it upon the shore. The roll was called by the captain, and thirty-three answered to their names; five were numbered with the dead.

The few hours of the past had been full of painful and distressing interest. The majority of our number had been mercifully rescued; but we were cast shelterless, with a small supply of provisions, with no clothing, only what was upon our backs, upon the most barren and desolate region of the earth.

What were our present prospects? They were dark and ominous indeed. A new voyage, in effect, was just opening before us, with diminished numbers, of the progress and termination of which we could not even entertain a reasonable conjecture; yet one thing was certain — its commencement was inauspicious. And, though hope might measurably sustain our minds, still the prospective view before this company of castaway seamen — the rigors of the arctic winter before us, wholly unprepared with clothing to withstand the merciless and long-continued cold of the north, uncertain whether there would be any deliverance for us by any friendly sail, or what would be our reception among the natives, — indeed, the prospect before us was any thing but cheering and encouraging.

But here we were, in the providence of God, vessel and boats gone, at an unknown distance from civilized life and from the settlements of the natives; this was our present lot. Self-preservation, therefore, prompted us to make immediate efforts, in anticipation of what we might need in the future. A common misfortune united all our interests and exertions.

The captain ordered that every thing of value to them in their present circumstances found among the wreck — such as provisions, casks of sails, pieces of canvas, ropes, broken spars, tools, whale gearing, &c. — should be selected, and brought out of the reach of the surf and the accumulation of ice upon the shore. More than a thousand barrels of oil had drifted ashore, and could have been saved had some vessel arrived about that time. A temporary tent was erected as soon as possible, in which various articles could be stored, as well as afford some protection to us from the inclemency of the weather.

There were two circumstances exceedingly favorable in our disaster. It might have been much worse, and no one might have lived to relate the sad event. We realized, upon the review, that this would have been our certain fate, had the ship gone ashore in the night time. It was, however, daylight, and thus we had a clear view of our condition, danger, and prospects. Had it been otherwise, and the same general features of the wreck been transferred to the darkness of night, we do not believe that one soul of us would have been saved.

The other favorable circumstance was, we were not cast upon a rocky part of the coast, or against some high and precipitous cliffs, which lift their bold and defiant fronts against the surges of the ocean far into deep water; to strike against such as we saw, would, at the first concussion, have been the last of the ship and of all on board.

In the good providence of God, however, we drifted upon a part of the coast which presented, for half of a mile or more, quite a plain, sandy beach. We were, therefore, wrecked in the most fortunate spot. On both sides of us, to the west and south-east, cliffs began to rise, and broken and abrupt ledges extended some distance into the sea. Though five of our number found a watery grave, yet the fact that so many of us reached the shore was a matter of profound gratitude to that God who controls the elements, and before whom the sparrow does not fall to the ground without his notice.

CHAPTER IV.

First Night on Shore. — Sleeping in empty Casks. — Parties of Exploration. — Dog Tracks. — Arrangements to leave the Wreck. — Desire to reach East Cape. — Reflections upon our Condition. — The dead Hog roasted. — The "pet Hog." — Company travel toward the South and East. — Two Natives seen. — Parley. — Directed to the Settlement. — The old Woman and her Ceremony. — The second Settlement. — Head Man cordial. — Men distributed among the Huts. — Not able to reach East Cape. — Company entertained. — Motives for it. — Government should reward the Natives.

THE first night we spent on shore was a very stormy one. There were rain, sleet, and high winds above and around us; below us, on the ground, ice, snow, and water in abundance. Our tent, which was a hasty and temporary construction, afforded us, after all, but little permanent shelter. The water came through and under it in every direction. Here we found an additional exposure, and the prospect of increased suffering both from the cold and wet. Had it not been for our oil, we could never have kindled a fire at first, nor continued it afterwards.

But necessity gives origin to many inventions and improvements. It suggests new plans, and urges to more favorable shifts and expedients. If, therefore, our arrangements for the first night's lodging on land should seem somewhat novel, or even unheard of before, let it be remembered that sad necessity drove us to this device.

If our frail tent with a few yards of torn sails stretched over us cannot shield us from the drenching rain, something else can. Most of us, on that sad and sorrowful night, got into empty casks; some were oil, others water or bread casks; it mattered not what, if we could only be protected from the violence of the storm, or rest in some place, instead of making the icy earth our bed.

With one head of the cask knocked out, and resting upon its bilge, one or two would get into each cask, and find within it quite a dry retreat. At the same time, a fire was kept burning not far from the open heads of the several casks, placed in a circle around the fire, and thus we were made as comfortable, perhaps, as our circumstances would permit. This was our first night's experience on land.

The next day, arrangements were made to form parties of exploration. We knew not where we were. Of this, however, we were quite certain — that we were north of the straits; but upon what part of the arctic coast we were cast away, we could not tell.

Therefore, our first object was, if possible, to ascertain our true position. The thick, foggy, and stormy weather which had prevailed for many days before the wreck, contributed greatly to mislead us. Neither sun, moon, nor stars had appeared for some time, by which we might have been guided in our course through the trackless deep. All above us had been shrouded with dense clouds, while strong and variable winds, approaching to the severity of tornadoes, and even hurricanes, had carried our ship far out and beyond her true course.

The last, and not the least perhaps, of the causes which resulted in our wreck, was the current, which appeared to change its course during the storm, as it not unfrequently does. At this time, it set from the eastern to the western shore of the Arctic. Under the combined influence of the current coming from the north and east, and a severe gale of wind, accompanied with hail, rain, and fog, our ship was constantly pressed upon the western shore, until she struck and went to pieces.

Two companies were now formed, one to take an easterly course, and the other a westerly one. Those who were left behind were to be employed in making sacks out of canvas, for the purpose of carrying bread and other provisions in our anticipated travels. This expedition was intended only as an introductory one to our final removal from the place, when we should ascertain more particularly where we were.

It was the opinion of some that we were cast away upon an island; and in so far as we could judge at this time, this opinion was rather confirmed.

The captain with his party took their departure towards the east, and Mr. Fisher and his party went towards the west.

Each man was armed with whatever defensive and offensive weapon he could well carry along with him. These weapons were neither guns nor swords, but a few knives, a hatchet, a broken whale lance, and a spade. We knew not with what we should meet, whether savage beasts, or more savage men.

The parties, as they traveled in opposite directions, soon found they were not upon an island, as they at first imagined, but merely upon an extended projection, as it appeared to be, from the main land.

The captain's party, after having traveled in an easterly direction about ten miles, discovered tracks of dog teams, and the footmarks of those who accompanied them. These facts assured the explorers that human habitations of some sort were not probably far distant. They therefore returned immediately to the tent to inform their companions, and to make particular arrangements for more extended researches. The western party also returned to the tent soon after the other, having made no very definite discoveries.

We became satisfied at this time, from the direction of the coast, and the general aspect of the country, that we were north of East Cape; but how far distant from it, we had no means of determining. It was likewise a matter of equal uncertainty whether we were east or west of East River. If we were west of this river, the prospect of our liberation the coming winter, or of meeting with any friendly sail, was extremely small. If, however, we were east of the river, we had strong hopes of deliverance before the winter should fairly set in, and that we should be able to reach East Cape in season to intercept some ship bound out of the ocean.

Thus we reasoned upon and discussed those matters which pertained to our speedy deliverance, or our bondage for months to come in the dreadful and merciless winter of the polar region; or it may be that no one of our number would escape to tell the wreck of the ship, and the catastrophe which would befall his fellow-companions.

Besides, considerations were urgent and pressing why we should make all possible haste, either to find some suitable habitation for the winter, or, perchance, fall in with some friendly vessel. With the advance of the season, we were assured that traveling would become more and more difficult, and that Borean storms would soon burst upon us with resistless fury. And hence, to remain where we were first cast upon the shore, without persevering efforts to save our lives, would be the height of presumption. With nothing more to protect us than the frail tent which we had erected, thinly clad, and all we had on our backs, a limited supply of provisions from the wreck, if we should remain at our first landing place until deliverance should come to us, then, indeed, before the opening of another spring, all of us would have fallen victims to inexorable death.

At this distance from the place where the scene of our sufferings commenced, how little can our readers appreciate what were then our condition, wants, and prospects! Indeed, ice and snow already began to largely increase, though we were in the region of eternal frosts, where they never wholly disappear. Both upon the shore, and as far as the eye could reach in an inland direction, the ice and snow were perennial occupants of the country. Neither the rains of spring nor the suns of summer are able to melt away and dissolve the deep foundations of a polar winter.

The surface of the country was much broken and uneven, and especially in the interior, alternating in valleys, deep gorges, precipitous cliffs, rugged and rocky eminences, one elevation rising above another, until the remote horizon exhibited lofty mountain ranges. The entire panoramic view presented an aspect at once sublime and frightful to behold.

It should be remembered that, amid our hopes and fears, we stood at this time on the borders of human habitations; and beyond this locality, as we afterwards ascertained, especially on this coast, there was but one known settlement of the natives to the north of us.

The dead hog that drifted ashore was skinned and roasted for supper on the second night after the wreck, and for breakfast next morning. Utensils for cooking were very scarce; only a few small articles had come ashore. The hog was suspended over the fire, and turned over and around when necessary, until it was baked suitable for eating.

Preparations were made on the coming day for another traveling-exploring expedition, in which all were to be included as one company. The grand purpose we now had in view was to find a passage to East Cape, or to fall in with the huts or settlements of the natives.

Before leaving, however, a grave question arose as to what should be done with the live hog, which had thus far shared with us in our deliverance from the wreck, and from his general deportment seemed to realize his forlorn condition.

This was a "pet hog" among the seamen; he knew his name, and appeared to have more than ordinary intelligence; at least, this was his reputation on board of the ship. His weight was not far from twelve score. He was washed from the deck at the time it was broken up by the sea, and discovered, by Mr. Fisher, floating about in the surf, and supposed to be dead. He went to him, and struck a smart blow upon his back, and said to him, "Jack, what are you doing here?" He immediately gave a grunt or two, started upon his feet, and struggled for the shore. He went with us to the tent, and made that his home. He would frequently wander forth some distance from the tent, apparently ruminating upon the sad state of things; and after a while he would return and take up his position at the entrance of the tent. In the estimation of the crew, he was indeed one of the "*learned* hogs."

When the question came up, what should be done with "Jack," many at once resolved they would never eat him, because he knew so much; and being so strongly attached to him, he really seemed like one of our number: "Old Jacky must not be eaten."

A different counsel finally prevailed. If we should leave him behind, he would soon be destroyed by wild beasts, and especially bears, that swarm the region; or he would perish with hunger. In view of these considerations, it was decided to kill the favorite hog, more from necessity than choice. He was accordingly killed and roasted, and a division made of him, each man taking his proportionate share. A cheese was also divided into as many parts as there were men, and distributed to each. Each man carried a sack containing thirty biscuits, in addition to other articles of provision just mentioned.

We were particular also before leaving, though it was quite uncertain whether we should ever see the spot again, to save from the wreck and stow away in the tent whatever we thought might be necessary for future use. This was a judicious precaution. As a last resort, if all other sources of deliverance should fail us, neither finding the settlements of the natives, or being received by them, nor seeing any ship to take us off, then we must return to the wreck, and make the best of what we had, and live as long as we could.

We furthermore agreed to travel seven days from the wreck, and if we found no help or deliverance from any quarter, then we would return, which would require seven days more—about as long as we supposed our provisions would last us.

The direction we took was towards the east and south, along shore, which was less difficult to travel than farther back in the country; besides, there were less snow and ice on the seaboard at that time.

We had traveled, as was supposed, about fifteen miles, when we saw two natives, some little distance before us, in an inland direction. At first they were unwilling to stop, probably aware from our appearance that we were foreigners. While we all kept together the natives continued on their way.

Captain Norton and two of his officers separated themselves from the rest of the company, making signs, thus indicating peaceable intentions, and advanced towards them. The natives then stopped. The captain and those with him approached them and shook hands with them. The natives appeared to understand the signals and signs, and at once desired that all the company that was behind some distance would come forward to them. This they did. The natives pointed in the direction of their settlement, and furthermore desired all the company to follow them. We followed them until we came in sight of their huts. Here the whole company was requested to stop, with the exception of the captain and two of his officers. We went with the natives into the settlement, and were immediately conducted into the presence of a very old woman, who marked one side of our faces with two lines, and our hands in the same manner, with a burnt stick. After this singular manœuvre was over, she made signs to the captain to call all his men, and they also were marked upon their faces and hands.

It is altogether probable that the marking of our faces and hands by this old woman with a burnt stick was some sacred rite, and that she might have been a sort of priestess or prophetess among the natives, and that the ceremony was a mark of her approval, or that she secured the protection of some divinity in our behalf.

It was ascertained afterwards, that this old

woman was held in very high repute among the natives, and that she was supposed to be a personification of a certain deity which inhabited some remote mountain in the interior of the country.

We also learned that the purpose she had in view in marking our faces and hands, was, that we might not *poison* those with whom we should eat, or contaminate any thing we should take hold of with our hands.

We were distributed among the several huts, and remained there that night. The natives set before us something to eat in the form of whale and walrus blubber, and deer meat. This "bill of fare" had not the recommendation of being cooked, but in its original state, with no other condiment than what age imparted to it. But whether the whole company found their appetites or necessities such as to pass immediately into this new regimen, was quite, if not altogether, improbable.

This settlement appeared to be of a temporary character; the natives with their families having come from another region or section of the country for the purpose of trading and hunting. There were but five huts in all. Our company, therefore, of thirty-three persons, occupied all the room they had to spare. It was close stowage but far better for us than to be exposed and unprotected during a long and chilly night.

After our arrival at this settlement, and some time during the night, word was sent by the natives, as we afterwards learned, to another and larger settlement, to inform the natives there that a company of shipwrecked mariners (*raumkidlins*) had come, and wanted shelter.

Accordingly next morning, ten or fifteen dog teams, with their drivers, made their appearance, having come from a settlement east of us for the purpose of transporting us, with our effects, to new and larger quarters. We arrived at this latter place about four o'clock in the afternoon, distance about twenty miles.

The captain, with two others, went directly to the head man of the settlement, whose name was *Taunty*, and made him understand, by signs and gestures, that they wished him to take care of the whole company.

He readily assented to our request. He manifested a most kind and obliging disposition. He showed a degree of sympathy for us in our destitute and dependent condition wholly unlooked for, and altogether unexpected. Such accommodations as he and his people had were promptly offered to us.

In this instance of cordial reception by the

NATIVE COSTUME.

natives, the hand of a good and merciful Providence can be easily discovered. What if, at this time of our need, the natives had thrust us away from their dwellings, and refused us shelter for the night, or a protection from the storm? or if they had exhibited towards us the spirit of hostility and war? Augmented sufferings would have been added to our otherwise unhappy lot. There would have been no escape for us from the arctic region. But we found friends when we most needed them.

We were distributed among the natives in the following manner: four men and one officer were to constitute a company; and in this proportion we occupied our respective huts, lived with the families, and shared in their accommodations.

Compared with the first settlement, where we stopped for the first night in our travel, this one was quite respectable, numbering twenty or more huts.

We had no intention of making this settlement a permanent resting place for the winter, if by any means we could find a more southern locality. We cherished strong hope of being able to reach East Cape, and thus being taken off by some ship passing through the straits, before the approach of winter. Nor was there any time to be lost towards completing such an arrangement as this.

One of the first things which we did, was to make known our wants to the head man of the settlement. So far as we were able, we conversed with him by signs, and thus endeavored to explain to him what we wished to do. He gave us to understand that it was impossible for us to travel down to East Cape this season of the year, and that the distance to the cape was very great, and it was therefore impossible to get there.

Not knowing our precise locality upon the coast, we could not tell whether the cape was three or five hundred miles from us. We concluded it would be safer to remain where we were than to venture upon such an uncertainty. It was afterwards ascertained that we were distant from East Cape about two hundred and fifty miles.

The head man gave us to understand, in his way, that there was a very great river to cross before we could get to East Cape, and that it could not be crossed now; and still further, if we should perish on the way, great ships (*laloutoutlines*) would come, kill him, and destroy all their huts.

On the whole, we judged that it was the desire of the head man that we should remain with him and his people, and live among them for the present; and nothing occurred in all our subsequent acquaintance with the natives in this settlement to remove this impression from our minds. It may be, however, that they anticipated some remuneration for their attention to us, which, by the way, they had a right to expect. This was not unlikely a motive which induced them to desire that we might live with them.

We sincerely hope the time may speedily come, when they shall be amply recompensed by our government for their kindness towards *thirty-three American seamen*, whom they protected, clothed, and fed, during three quarters of a year.

CHAPTER V.

No Prospect of reaching East Cape. — Painful Conviction. — The Province of Christian Faith. — The Wreck visited. — The Natives. — Hope unexpectedly revived. — Ship in Sight. — Comes near. — Signals from the Land. — No Assistance offered. — Sails down the Coast. — Indescribable State of our Minds. — Card in The Polynesian.

The prospect of reaching East Cape for the present was at length abandoned. A conclusion was arrived at, from the necessity of our condition, which was full of disappointed hope, and which required an unusual degree of patient courage to sustain our minds under the painful conviction that we must, after all, spend the next three quarters of a year, if we should live, in the northern regions. How the mind of man becomes shaped and adjusted to meet certain conditions of his being! If viewed in the light of unavoidable necessity, we see the force and independence of mind grappling with adverse circumstances, thus proving its original superiority over all outward disadvantages. It is, however, the province of *Christian* faith in the providences of an all-wise God, which secures to the mind true reconciliation, imparts hope in adversity, and awakens unearthly joy in seasons of sorrow and disappointment.

The next day after our arrival at our new habitations, the whole company rested, and got somewhat recruited as to our bodies, and, not the least in our circumstances of anticipated captivity for months, our minds became partially settled that we must make the best of a common disaster and a common destiny.

The day following, preparations were made by ourselves, in connection with the natives and their dog teams, to visit the wreck. One of the first questions asked, and the principal one in which the natives were more interested than in any other, was, whether there was any rum at the wreck. A keg of spirits had been washed ashore, as before stated, and a part of it had been used, and the remainder was in the keg in the tent, stowed away with other articles from the wreck.

A difficulty was now apprehended. If the natives should find the keg of rum, and become intoxicated, as they probably would, serious and perhaps fatal consequences might take place. To avoid any fears of this sort, and remove all grounds of contention, the captain sent two of his men ahead, with orders to knock in the head of the rum keg. It was done as commanded; no further difficulty, therefore, could arise from this source.

Self-preservation prompted to this; but in a multitude of instances no less striking, where property, reputation, and even life itself are concerned, a like decision, to knock in the head of the rum keg, or break jugs and bottles, and pour the source of evil upon the ground, would be highly commendable, and fraught with the most happy results.

In due time we reached the wreck, and, as was expected, the natives began to search for spirits; but for their advantage, as well as ours, they found none. They sought every where for it, ransacked every nook and corner, hauled over wreck stuff, looked into barrels, knocked to pieces oil casks, &c., to find it, but all in vain.

It appeared, furthermore, as if the natives supposed they had a *right* to whatever they could lay their hands upon, and what they found among the wreck, or on shore, was a lawful prize. Several pieces of white and blue cotton cloth had washed ashore since the wreck was last visited; these the natives appropriated to their own use. A slate was found, and upon it we wrote the name of the ship, her captain, and where the crew could be found, and placed it in a prominent position near the wreck, hoping that it might possibly meet the eye of some deliverer, though an event so much desired could now hardly be expected.

The company remained in the vicinity of the wreck until towards night, and then each man took with him a bag of bread, and, with the natives and their dog teams, we left for the settlement, which was about fifteen miles distant.

It was exceedingly hard to visit the scene of our recent disaster, and behold the desolation and end of the noble ship that had withstood so many storms and weathered so many gales, but now a promiscuous mass of broken timbers, planks, and spars; besides, her cargo thrown upon the beach. If possible, it was even harder to leave what remained of her behind, and to carry away a small quantity of provisions to eke out an existence which, under the most favorable circumstances, among the natives, must be most trying and painful. And then, again, all the provisions we expected to obtain from the wreck could last us but a few months, at the longest. If our lives, therefore, should be prolonged, we saw before us the only alternative of living as the natives did, being constant spectators of their extreme filthiness in person and habits, and sharing with them in the peculiarly offensive and disgusting character and preparation of their food.

The next day, the company remained in the settlement, wearied with the labor of the preceding day, and, the greatest calamity of all, oppressed in our minds, as we contemplated the future; and as we began to realize more and more what would probably be our destination for many long months to come.

"Hope deferred maketh the heart sick;" but hope revived — when well nigh abandoned and ready to expire, like the last flickerings of the lamp — hope revived imparts new life, and sends a thrill of joy through languishing minds. Thus the weak become strong, and the disheartened are animated and encouraged to put forth more earnest efforts. Hope revived under the circumstances in which these shipwrecked mariners were placed was like the introduction of light, comfort, and home into their wintry habitations. What intelligence more to be desired and sincerely asked for than the announcement of a sail in sight?

Think of them, as brooding over their anticipated doom; settling it, or having settled it in their minds, that their abode was doubtless fixed for the present; thoughts of home now and then rushing into their minds with overwhelming force, or, it may be, with the only exception of their sleeping moments, never out of their minds; indeed, their very dreams shaded, colored, and made treacherously illusive with joyous meetings of companions, parents, relatives, and friends! Think of them at such a time as this, when the hope of deliverance had taken its lowest dip, like the wintry sun of the Arctic passing below the horizon, its light and comfort quite departing; so hope in the minds of this company of wrecked mariners had fallen beyond any reasonable expectation of deliverance.

Severe and terrific storms of wind, rain, hail, and snow had swept over the northern ocean, and ere this it was supposed that every ship had sought a more southern and genial clime.

What, then, was our unexpected and glad surprise, on the following day, when, amid the tumult and confusion, as well as the excitement of the natives, both in and around the huts, it was announced that a *sail* was in sight!

With all possible speed we hastened to a high cliff bordering the sea shore, and there we saw, indeed, what our eyes delighted to behold, and our bosoms swelled with grateful emotions to contemplate — a ship under sail, some ten or twelve miles distant, and standing in directly for the shore. As we looked, never before with more exhilarated spirits and reviving hope, on, on the vessel came, approaching nearer and nearer, until her davits were plainly seen, and men walking to and fro on deck. The ship now was not more than two miles distant. She came to, main yard hauled back, and lay in that position a quarter of an hour or more.

With these indications, all doubt had nearly or quite left our minds that the intentions of those on board were to take us off. Still, no boat was lowered, nor was there any answering signal. This surely was mysterious, and betokened fear. And yet could it be that within so short a distance no deliverance would be extended? It was contrary to reason to believe so; the thought must not be cherished a single moment. We should soon tread a friendly deck, and share again a sailor's home on the deep. Thus whispered hope, suddenly revived in all our hearts.

But in order to make the case doubly sure, and remove all suspicion in the minds of those on board that those on shore were not all natives, two colors, one white and the other blue, were raised upon poles to the height of full thirty feet. It was plainly seen by those on board, as subsequent testimony from the officers abundantly proved. Besides, these were signals of civilization, of common brotherhood, of pressing emergency, and strongly excited hope. But, alas! they met with no response from that vessel's deck.

Lest there should be a lurking distrust in the minds of the captain and officers of the ship that these signals were a mere trickery or device of the natives to get on board of the ship, or for the ship to send a boat ashore, the company on shore separated themselves from the natives, so that with the aid of a glass, or even with the naked eye, a distinction in manner, movement, and dress could be easily seen by those on shipboard. This expedient also failed.

As another resort to attract attention, a fire was kindled; and yet the rising and curling smoke met with no cordial response, no friendly salutation; no boat came to our rescue. Shortly after, the ship filled away, passed down the coast, and was seen no more.

We felt, what no language can adequately express, that this was an instance of cold, deliberate, and even infamous neglect. Could it be they were ignorant of the ordinary laws of humanity, and wilfully misconstrued the most obvious signs of needy and suffering seamen? Instances have, indeed, occurred, in which vessels at sea have been known to pass near shipwrecked mariners, and yet they were not discovered. They were upon a low raft, perhaps, or had no means of raising a signal, and were therefore passed. The imploring cries and stretched-out hands of the sufferers were alike unheeded; not from any intentional neglect, by any means, but simply because they were not seen from the vessel's deck.

It is sad to contemplate an oversight even like this, in which the hopes and lives of a number of unfortunate seamen were suspended upon the bare possibility of being recognized by the passing ship.

How many, many have doubtless perished in mid ocean, whose eyes beheld again and again the approaching and departing sail, whose hearts alternately rose in hope and sunk in despondency, and yet at last died without the precious boon of deliverance!

Other instances have, however, occurred, of a far different character. Suffering, exhausted, and dying mariners, either upon wrecks or rafts, have been left uncared for and abandoned by the passing ship.

If the records of the past did not furnish conclusive evidence of the truth of the foregoing statement, it would seem that the bare announcement of the fact would be sufficient not only to appall the hardest heart, and cover with deep and lasting shame the perpetrators of such a deed, but to place it in the frightful category of those events absolutely beyond both human experience and credulity.

Revelation informs us that "the sea shall give up its dead;" so will there be a resurrection both of the good and bad in human conduct. A virtuous and benevolent act performed upon the ocean will never be concealed. The winds, as they sweep over its surface, will declare it. And so, on the other hand, an act of inhumanity, capriciousness, cruelty, or turning a deaf ear to the expostulations and entreaties of the dependent and suffering, will never slumber. The mighty waves, as they traverse the great deep, will speak in thunder tones that the deed lives.

The hopes of *thirty-three* persons in the cold and dreary region of the north, in the province of perpetual ice and snow, were suddenly and unexpectedly revived by the near approach of a ship within trumpet hail; signals of wrecked mariners on shore, the ship remaining more than fifteen minutes with her yards back, and those on board beholding the demonstrations of intense anxiety of those on shore that deliverance might be sent to them, and yet not one motion made for our rescue! The ship is soon on her way, and out of sight. If hope was ever suddenly and unexpectedly revived, it was then; if hope was ever suddenly cast down to its lowest depths, it was then.

Nor could our eyes hardly believe what we were beholding. Was it all illusion, dream, or magic? No; it was a reality. We had been tantalized. The cup of the greatest earthly blessing had been held to our lips, and yet we were not allowed to drink of it, but it was dashed to the earth in our very presence. The departure of that ship was the departure of mercies to us, to procure which we would have been willing to make the greatest earthly sacrifice.

What a day of joy and sorrow was that to us! How many hitherto downcast countenances were lighted up! What words of good cheer passed from one to another! How many hearts bounded with thankfulness and gratitude at the thought of so speedy a deliverance!

Our families and friends at home were thus far ignorant of the distressing scenes through which we had passed, and also of our present condition; but ere long, as we believed, on our arrival at the islands, we should communicate to them the wreck of our ship, the loss of the voyage, and the fortunate rescue of so many of our number from a watery grave.

We felt that we had much for which to be thankful to God, and that soon we should be able to send to anxious ones at home the happy intelligence that we were among the saved.

Such is hope when strongly excited. It ennobles and invigorates the human soul; it adorns the horizon with the gorgeous drapery of morning clouds; it paints the evening with the glories of departing day; it forgets the past; it is the elixir of life itself; without it man lives only in the present, and anticipates no future good.

But that was a day of sorrow too! It seemed as if we should sink into the very earth, and that we were unable to stand, with such a load and pressure upon our spirits. We were crushed both in body and mind. Contending emotions of indignation, abandoned hope, unmitigated grief, and poignant sorrow, swayed and strongly agitated every bosom. The whole company wept like children.

It may be asked, "Why did not the officers and crew avail themselves of the canoes of the natives, and go off to the ship?" It is true there were several canoes near the shore, but the natives were unwilling they should be touched; from what cause we could not understand. Our acquaintance with them, and theirs with us, had thus far been very slight; and it may be they had serious suspicions in their minds that we designed some evil towards them. They were doubtless governed by some notions, in refusing us the aid of their canoes, in keeping with their half-civilized or barbarous natures.

The captain and others offered to hire the canoes, at the same time presenting to them some little articles they had with them, as a pocket or jack knife, but all to no purpose. They resisted every proposition.

The officers and some of the crew were so anxious to get to the ship that they proposed twice to the captain to take forcible possession of the canoes, and follow the ship; and they would have done it, and risked all the consequences, had the captain approved of it. He, however, opposed this plan, on the ground that though a few might succeed in reaching the ship, yet those who were left behind, being entirely unarmed, would probably be instantly killed, and, therefore, it was bad policy to expose the lives of a majority of the company for the safety of only a few. Or, it may be, in their first efforts to seize the canoes, and before they could even get them into the water, the natives would fall upon us, and massacre the whole company on the spot. And still further, we were wholly in their power, both for the present and for months to come, and without their kindness and good will we had no sort of chance for life; therefore the least misunderstanding or violent collision between the parties might lay the foundation for causes which would result, if not now, yet in some future time, in the destruction of the whole company. These considerations, suggested by the captain, dissuaded his men from attempting a forcible seizure of the canoes of the natives; and, therefore, for the good of the whole, that means whereby a few possibly might have reached the ship, was given up.

We leave this painful reminiscence of the past by copying from *The Polynesian*, published at Honolulu, November 19, 1853, the following

CARD.

"The undersigned, late master of the whale ship Citizen, of New Bedford, feels it a duty he owes alike to the living and the dead to make known the following circumstances.

"On the 25th of September, 1852, in the Arctic Ocean, in lat. 68° 10′ N., the ship Citizen was wrecked, and five men were lost; himself and the balance of the crew reached the shore, without any thing but the clothes they stood in. It was very cold, and they kept alive by burning casks of oil that had floated ashore from the wreck;

that they lived near the wreck until October 3, when the whale ship Citizen, of Nantucket, Captain Bailey, hove in sight; they immediately hoisted a flag upon a pole thirty feet high, and made every signal they could of distress; that the ship at first stood in as though she saw them, then hauled up and shivered in the wind, and afterwards filled away and left them. She was so close at one time that those on shore could see her davits. The feelings with which they saw the vessel leave them are indescribable, as no hope was left them but to endure the rigors of a winter's residence in that cold, bleak, and desolate region, if they should escape the tomahawk of the savage. That their signals were seen by Captain Bailey there can be no doubt, as Captain B. reported seeing his signals last fall. The mate of Captain Bailey's vessel reported to Captain B. that he could see sailors on shore, and requested a boat to go to their relief, which Captain B. refused.

"Through the inhumanity of Captain Bailey, we were compelled to remain *nine months* in that barren region, destitute of clothing and food, other than the natives could supply us from their scanty stores of blubber and furs. During this time, two of the crew perished from cold, and left their bones to bleach among the snows of

Hunting the Polar Bear.

the north, as a monument of 'man's inhumanity to man.'

"The natives were humane, kind, and hospitable to us, though wretchedly poor.

THOMAS H. NORTON."

CHAPTER VI.

Our sad and desolate Feelings after the Departure of the Ship. — What we should soon witness of Arctic Winter. — The Wreck visited from Time to Time. — Provisions transported to the Settlement. — The Weather. — Whales near Shore. — Severe Gale of Wind. — Fall of Snow. — Ocean frozen over. — Sudden Introduction of Winter, and its Dreariness. — Not to be described. — The Sun falling, Nights lengthening. — Disappearance of the Sun. — Long Night. — How we passed our Time. — Confined to the Huts. — Singing. — Neither Book nor Chart, nor Writing Materials, except Pieces of Copper. — Hope of Liberation another Year. — Captain Norton's Method of keeping Time. — The Razor. — Our Clothing. — Provisions getting low. — Natives both eating and stealing ours. — A new Chapter. — Commenced living on Blubber with the Natives. — Native Stock diminishing. — Winters in the Arctic vary. — The native Manner of capturing the Whale. — Preparing their Food. — Native Bread. — Description of their Huts. — Their peculiar Locality. — Their Method of lighting and warming them. — The Filthiness of the Natives.

The next day after the departure of the ship, as well as the departure of our highest earthly hopes, — hopes which had been excited in us immeasurably beyond any former experience, — we remained principally in our huts, having neither desire nor energy, heart nor hope, to go abroad, but what was most fitting in our present condition, and future prospects, to indulge in sad and melancholy reflections upon the few past hours.

There was a singular solitariness pervading all our minds, such as we never felt before. We were now painfully sensible that the ice, snow, and cold, peculiar to this region of the north, such as we never witnessed before, would ere long form around us an impassable barrier, frightful even to contemplate, and through which there would be to us no present egress.

What remained of provisions still at the wreck, and other articles which may have washed ashore, reason and the instinct of self-preservation taught us, it was our duty at once to secure. Accordingly, the day following, the whole company were again assembled, and went to the wreck. We made a division of the provisions, especially of bread, between the different parties occupying different huts, and each party transported its respective share to the settlement. The natives were present with the crew during the day, and ever ready to appropriate to their own benefit whatever they saw fit to take, or were disposed to lay hold of.

There were several casks of molasses which came ashore; and since these could not be very well divided at the wreck, it was resolved to construct a species of sled, upon which a whole cask could be drawn to the settlement at one time. This we did, though it required many tedious hours and severe labor. By the aid of ropes, and a combination of all our efforts, we succeeded in getting all the molasses to the huts. We managed in the same way with a number of barrels of flour which came ashore about this time. Several tin plates and basins were also found on the beach, and these answered an excellent purpose, as they afterwards proved; because in them we mixed our flour and molasses together, and thus made very luscious pancakes. We usually baked them outside of the huts, as no fire was allowed within, except very rarely; nor were we permitted to make any outside when the wind was in a northerly direction, lest the smoke should frighten away the seals from the shore and region. So reasoned the natives.

We continued to visit the wreck and obtain whatever we could, until the weather became so severe, and the traveling so bad, that it was no longer safe to expose ourselves.

After having gathered all the provisions we could find at the wreck, such as bread, flour, and molasses, we judged that with economy, and with ordinary allowance, it would last the ship's company three or four months. But the great drawback which we apprehended, and which we found to be true, was, the natives acted as if they had as good a right to our provisions as we had ourselves. They not only joined us in eating what belonged to us, but they took what they wanted, both openly and secretly.

The weather continued quite moderate, we should judge, for this region of the north, not intensely cold, still gradually increasing, until the 17th of November. While the sea was open, whales were very plenty. They came near the shore where our settlement was located, and sported among the breakers, and in some instances, would rest their huge heads upon the rocks, just on the surface of the water.

About this time, a very severe gale of wind blew from the north, more furious and winterish than had occurred since our abode in this region, accompanied with a heavy fall of snow. The wind was so violent that it prostrated several native huts. This storm was doubtless a forerunner of winter indeed, and which brought from the remote wastes of the Northern Sea vast quantities of ice, which, in connection with that which had been forming along the coast, closed up the whole ocean as far as the eye could reach. Indeed, all water entirely disappeared.

This was an uncommon and singular feature in our experience of an arctic winter. It thus began in earnest to put on the sterner and more terrible attributes of dreariness and desolation.

There was something profoundly dreadful and awe-inspiring in the giant march of a polar winter, prodigious in its increase of snow and the vast accumulation of ice. It was upon a scale of operation so sublime and awful as to baffle all human description, and throw wholly into the shade, as absolutely insignificant, the intensest winter ever experienced in our native country. It is utterly impossible to give to any one who has not shared somewhat in the tremendous reality of the scene a just conception of it.

The sun was now falling rapidly, and showing its bright disk only a few hours above the horizon. The nights were very long, and the days were becoming shorter and shorter. It seemed as if the luminary of day was indisposed to throw abroad his own rays upon a region of the earth's surface where either human or animal life could with so much difficulty exist. In a few weeks, the sun had wholly disappeared, though his track of light could be distinctly traced in his course below a section of the horizon; but still it was becoming fainter and fainter, until total darkness and a long night of nearly a month enveloped the outward world, as well as enshrouded our own minds in indescribable gloom and sadness.

Our readers may inquire how we passed our

time during our detention among the natives, and especially during the coldest of the weather, or during the long night of polar darkness. When the thermometer, during the depth of winter, would doubtless have indicated scores of degrees less than zero, we rarely ventured forth out of the huts. But far otherwise with the natives. They would go out and travel from settlement to settlement, even in the coldest weather. At times, however, they would return from their winter excursions somewhat frost-bitten. We also became, in a measure, accustomed to the intense cold, and being clothed in the garments of the natives, consisting wholly of skins and furs, we could endure a great degree of cold.

If there was any outward relief to be found to our minds during the long nights of the arctic, and the entire absence of the sun for several weeks, it consisted in the peculiar and uncommon brilliancy which marked the course of the moon in those clear skies.

Nor was this all. The aurora borealis there is seen in all its native beauty and grandeur. It illumined the sky with a light but little inferior to that of moonlight. It would from time to time shoot up and spread itself over the whole northern horizon, and with its sparkling scintillations and brightly-colored coruscations, it would form a splendid arch over our heads. And then again, as the advancing column of warriors rushes into battle, so the bright line above us, with its moving front and wheeling battalions, would seem to change its hue and position, and thus prepare for a fresh onset.

The aurora borealis of the arctic and polar region is one of nature's grandest and most sublime scenes ever beheld by mortals.

As we were confined within the huts of the natives during a greater part of our abode with them, and as nothing particular occurred demanding our exposure out of doors, we had sufficient time to sleep, if sleep we could. To pass away time was extremely hard and irksome. Its wheels rolled slowly and heavily along. Some of us would sing to the natives, which tended not only to divert and encourage our own minds, but to please them. We found, however, they were wonderfully pleased with our singing, and so much interested were they in it, that nothing would satisfy them unless some one of us was singing to them. Thus they laid an oppressive task upon us, which we were not able to perform. What we commenced, therefore, as a sort of pastime, in order to while away tedious hours, days, and months, finally became, through the constant importunity of the natives, a grievous burden to us.

We had neither book nor chart of any description in our possession, with which to divert or instruct our minds. We had nothing upon which to write any event or fact, except small pieces of copper, and a few stray leaves which we happened to find in the huts of the natives. Our time, as all must see, was spent comparatively in a most listless and unprofitable manner; it was simply the endurance of life, and the prolonged hope that another year, if we should live to see it, would bring to us the day of deliverance.

Captain Norton kept, by the aid of a piece of twine, in which he tied knots, an account of every day, from the time of the wreck until our rescue at East Cape, with the single exception of only one knot too many, which he supposed he must have added during the long night.

The only razor, which was a great favorite with the company, and which we frequently used to the best of our ability, without either soap or brush, was an ordinary jackknife. It was necessary to keep our beards trimmed within proper limits; otherwise our breath, even in the huts, and especially when exposed to the air outside, would reduce them to a mass of solid ice.

These two articles, viz., the twine and the knife, were about all the significant and expressive mementoes which we brought with us from our arctic quarters. These, however, were sufficient to bring most distinctly and vividly to our minds a painful episode in our ocean life.

The clothing with which we were furnished by the natives, and without which we must have perished, was composed of skins and furs. We dressed as the natives did. An observer could have seen no difference in this respect between us and them. Our shoes, pants, and a kind of jacket, and caps, were wholly of skins, with the hair inside, and then over these another dress, with the hair outside. Thus clothed, we were protected from the keen, piercing air — a protection secured to us which no other substitute could provide.

About the 1st of January, in the depth of winter, we began to perceive that nearly all the provisions we obtained from the wreck were about gone. The natives had shared with us in the several huts to a considerable extent in consuming what belonged to us. They were very fond of our flour, molasses, and bread. They wanted to eat what we ate, and when they could not get it by fair play, they would indulge in their natural propensity, and steal it.

What we greatly feared was now coming upon us. A new chapter in our history began to open. The food of the natives must henceforth be for our support. To their credit, however, be it said, there appeared no disposition on their part to confine us down to a mere pittance, while they themselves had their usual allowance. What they had was freely offered to us, and both parties fared about the same while food lasted. Their supply of provisions for the winter, so far as we could judge, was not large; but now the addition of thirty-three persons to their number soon diminished their usual stock.

Winters even in the arctic are variable, as we learned from the natives; some were very severe, and other less so. We ascertained that an entire settlement to the north of the one in which we lived, and north of the wreck, perished a few years before in consequence of the intense cold, and the want of provisions.

Some idea perhaps may be entertained by the reader of the principal kind of food the natives eat, and what we lived upon for months while with them.

THEIR MANNER OF CAPTURING THE WHALE.

As the whale approaches quite near the shore, the natives are not greatly exposed by following him to a great distance in their canoes. They take their own time and opportunity for killing the whale. Both men and women are in the canoes on such occasions. It is regarded by them as a family affair.

They go sufficiently near the whale to throw a harpoon into his body. Their harpoons are somewhat different from ours, yet in principle they are precisely the same. When they have thrown one or two irons into the whale, they cast overboard two air-tight and inflated seal skins attached to the lines. Every canoe in pursuit of whales has two skins of this sort. If the whale is disposed to turn flukes and go down, he must of course carry with him these full blown skins. The lines are very strong, being made of walrus skin. When the whale makes his appearance again, he is struck by another canoe, and two more seal skins are attached to his body. Thus they go on fastening irons into his body, and impeding his course by any number of seal skins, until he is wearied out, and then they go up to him and lance him. Whales have been picked up, by ships and boats, having several seal skins attached to them. One whale was found, several years since, which had *twenty-eight* full-blown seal skins trailing after him.

PREPARING THEIR FOOD.

Having captured the whale and drawn him ashore, they then proceed to the work of cutting him up, and stowing him away for future use. Both the blubber and entrails are deposited in a place together, especially prepared for the purpose. The place is a circular cavity, in the form of a cellar under ground, from five to eight feet deep, and with varied diameter, from three to five feet. These depositories are placed along shore, some distance apart, most convenient for receiving the whale when taken. In this cellar they deposit not only the whale blubber and its intestines, but also the blubber of the walrus and seal, and occasionally a deer is thrown in, with all that appertains to it, except its skin and perhaps its feet. The whole is thus mingled together in due proportions, and eaten by the natives with no further change in the promiscuous and offensive elements than what time itself would produce.

It was from such storehouses as these that the natives drew out their chief support for the winter, and nearly for the whole year. It is quite impossible to define this compound, and even if we could it would answer no good purpose; for with us it is profoundly obnoxious even to think of it. Absolute necessity — the simple fact of an existence, compelled us to live upon such qualities of food, compared with which our hogs have dainties, and luxuriate upon the fat of the land.

There were, however, some exceptions. Now and then a fresh seal was caught, a bear, a walrus, or a deer brought in. But whatever other good qualities the food might have had, it was all eaten raw; at least this was the case with the more northern settlements, and particularly the one in which we spent most of the winter. The natives farther south exhibited some slight improvement in the manner of preparing their food; yet on the whole, the difference was very small, and not worth mentioning.

At the time of our meals, if they can be called such, all the members of the hut would gather around a large dish, or tray, or trough, as much like the ordinary hog's trough as it could well be, and then each one would either help himself to what there was in it, with his hands and fingers, or receive his piece of blubber from the head man of the hut. In this manner the natives took their meals. From the necessity of our condition we had to conform in a measure to this foul and disgusting custom, to say nothing of the filthy nature of the food; we were compelled to eat or starve.

The bread used among the natives was made by boiling a vine, which they find on the ground in those places where the snow melts off during summer. This vine is somewhat bitter. They make a practice of collecting it during the summer months. After it is thoroughly boiled, they pack it away in seal skins for future use. This is all the bread they have.

HUTS.

A brief description of the huts of the natives may not be out of place in this connection.

The huts are generally round, differing in size in proportion to the family, and averaging, perhaps, from twelve to thirty feet in diameter. The lower part of the hut, and to the height of four or five feet, is well secured with upright stakes, situated a few feet apart, and fastened to each other by cords of walrus skin. The huts, and especially those where we were located during the winter, were not made partly underground, as was the case with some we saw in the direction of East Cape, but so constructed on the surface of the ground as to be easily taken down and removed.

From the lower, or upright part, the roof extended in an oval form to the height of ten or twelve feet. At the termination of the top, or apex, there is an opening, which is closed or otherwise, according to the state of the weather. This opening affords about all the egress to the smoke of the lamps and fire, when made in the huts. It is very rarely, however, that fire is made in the huts.

The covering of the huts is usually of walrus skins, and impermeable to water. There is generally but one door to the hut, which is somewhat smaller than ordinary doorways.

The interior of the hut is divided into two principal rooms, or apartments, one of which may be called the eating room, and the other the sleeping room. The sleeping apartment is separated from the other by a temporary screen, which can be easily drawn aside or gathered up. The sleeping apartment is again subdivided into smaller sections, to suit the convenience of the family. The partitions are of walrus or deer skins, as a matter of course. These rooms are much warmer than one would naturally expect to find in this cold region of country.

The bedsteads (so to speak) are the skins of walruses, stretched upon and fastened to the tops of stakes about one foot from the ground, under which a bedding of coarse rushes is placed.

The pillow, or that upon which the head may rest, is made by drawing the walrus skin over one end of a stick, or log.

The peculiar *locality* of the huts or settlements is another consideration deserving a passing notice. The natives select the bleakest spot in the region for their settlements, where the wind blows without any obstruction. They, therefore, avoid all shelter behind hills, or cliffs, or in valleys. In placing their huts in such exposed localities, as for example, upon a plain, or level, near the sea shore, their purpose is to secure protection from the drifting snow, which otherwise, were they in the lee of some hill, or rising ground, or in a valley, would cover them up, and overwhelm them. Besides, the huts being circular, the wind and snow have opportunity of circulating in such a manner as generally to leave a clear space of several feet or more around the hut.

Notwithstanding all the precautions of the natives to avoid the drifting snow, still it was so deep at times upon a level, that when passing along, and even quite near the huts, we could not discern the tops of them, and should not have known that we were in their immediate vicinity, had it not been for tracks we discovered in the snow, or from the barking of the dogs.

THE METHOD OF LIGHTING THEIR HUTS.

The lamps are in the form of a hollow, circular dish, somewhat in the shape of a bowl, made of clay. This vessel is filled with seal's blubber, and around the edge of the lamps inside, is placed a row of moss of fine quality, obtained from the mountains. This moss is set on fire, and by its heat the blubber in the vessel is converted into oil, which in turn feeds the moss, and thus good light is obtained. Two or three such lamps in a hut would afford considerable heat.

The smoke, however, which proceeds from them is immense, and exceedingly offensive. It is so thick, that every article in the hut is covered and blackened with it. When one comes to clear air and breathes, there will be seen a volume of darkened vapor going forth from his nostrils and mouth.

These lamps are burning nearly all the time, and especially when the days are short, and during the long night of darkness in midwinter.

Neither the smoke from the lamps, nor the quality of food we had to eat, nor the manner of eating it, nor constantly observing the filthy habits of the natives, was all the degradation we felt and experienced.

With the strictest propriety it can be said the natives were loaded with vermin; and yet as indifferent, apparently, to such a condition, as if it were the most trivial circumstance in the world. Indeed, they appeared to enjoy the presence of the innumerable hosts that swarmed in all parts of their huts. Their persons, garments, skins in the huts, sleeping apartments, &c., were literally alive with them. The misery of such a state we have neither words nor heart to attempt to describe.

POLAR BEARS.

CHAPTER VII.

Health of the Natives. — Their Diseases. — Captain N. prescribes a Remedy. — Their superstitious Notions. — Mr. Osborn prescribes for the Sick. — A fatal Case. — They surround Mr. O. with threatening Gestures. — Native Remedy for Nose Bleeding and Sore Eyes. — Burial Ceremony. — Marriages. — General Appearance of the Natives. — Their Character. — Their Habits of Industry. — Property. — Language. — Icebergs. — Their Formation. — The Distance to which Icebergs float. — Their Magnitude. — Field Ice. — The sudden Disappearance of Ice. — How accounted for. — Icy Vapor. — Poisoning.

THE HEALTH OF THE NATIVES.

So far as we could learn, they had the usual share of health with other communities. A good proportion of them reached an advanced age in life; and some, we should judge from their appearance, were much older than the oldest among our own countrymen.

DISEASES.

One of the most common diseases among the natives appeared to be that of worms — originating, probably, from the character of their diet. Their medicinal preparations were but few and simple.

Captain Norton had in his possession a package of wormwood, which he picked up on the shore near the wreck. Though it had been saturated with salt water, yet, from time to time, he administered a strong drink of it for the above disease, with complete success. As proof of their appreciation of his services, they would put a dried crow's head upon his arm. His success went so far, that he had as many crows' heads as could be strung from his wrist to his elbow.

Captain Norton, however, was wisely cautious in one respect, and that was, he would not prescribe in any given case of sickness, unless he was well convinced there was no immediate danger to the patient, or that he could afford some temporary relief.

Their superstitious notions were such that, if any prescription should fail, and the patient should not recover, they would suppose at once that the proposed remedy was the cause of death. It required great prudence, therefore, to manage, not only the sick, but also those who were well.

A case occurred of a very trifling character at first, but finally it proved fatal. The face of a person, a woman, was somewhat swelled; the cause of it, so far as we could ascertain, originated in a defective tooth.

Mr. Osborn acted the part of a physician at

this time. He applied a poultice of sea bread, in order to reduce the inflammation, which he supposed it would shortly do.

The woman, however, did not get immediately better; and her friends took the poultice off, and in the place of it they tied a string very tight round her chin, in order, as they believed, to prevent the disease or swelling from going downwards. The string rather increased the inflammation; and then it was taken off, and placed still lower down, until the swelling had very much increased, and had reached her breast. The string was now tied tighter than ever, until it became embedded in the flesh.

Since Mr. Osborn's remedy had failed to benefit her, the natives, from their appearance and gestures, supposed that the poultice was an injury, instead of an advantage, to the woman. They therefore gathered round Mr. Osborn in the most threatening attitude, and he greatly feared they were about to injure or kill him. At any rate, he learned one important lesson — to be more cautious, in future, in prescribing remedies to the sick among the natives. They were superstitious, and therefore unreasonable.

The natives were subject to nose bleeding; the excessively cold weather was doubtless the chief cause of it. The remedy which they employed, and in use among them, was the application of a frosty stone, or piece of ice to the back of the neck.

Sore eyes were quite prevalent among them, more so in some seasons of the year than in others. This disease is caused by the reflection of sunlight upon an almost boundless surface of snow and ice. The simple remedy, in ordinary cases among them, was in making a slight incision with a thorn or some sharp instrument in the flesh, directly between the eyes, so as to draw several drops of blood. The effect of this treatment was to reduce the inflammation, and thus carry off the soreness from the eyes.

In some instances, however, the eyes of the natives had wholly run out.

BURIAL CEREMONY.

This, in many respects, was very peculiar, and quite different from the great majority of semi-barbarous or half-civilized tribes and nations. When one dies, a wife, for example, — as this instance did occur in one of the huts, — the following ceremonies were observed to take place: —

Immediately on the death of the person, or just before death took place, the relatives and friends gathered in the hut, and commenced a most bitter and vociferous wailing or mourning.

The usual means to expel the disease, whatever it might be, had been employed in vain. Several skins, stretched over hoops varying in size, had been broken by furious beating, accompanied with fantastic gesticulations and almost unearthly sounds, if possible, to cure the patient. But all to no purpose. They now found that death was approaching; and since every effort of theirs had not benefited her, they pronounced her incurable, and proceeded at once to terminate her existence. She was not permitted to die wholly from the natural effects of the disease; but a small cord was placed round her neck, and gradually drawn closer and closer by those who stood on each side of her, until life became extinct. During the last scene, she gave various presents to her relatives and friends. She died with singular indifference, and without a groan.

Whether all the sick, who, they supposed, would not recover, were thus put to death, as in the foregoing instance, may be a question. Yet, in so far as could be ascertained from observation and from conversation with the natives, it is the opinion of those who lived with them for several months that this was generally the case.

Soon after, all the remaining property which she possessed — her clothing, needles, combs, beads, &c., besides some tobacco — was sewed up with her in the dress she usually wore, or in which she died. A new sled was then made for the deceased, and two of the best dogs in the family were selected to bear away the corpse. Instead of carrying the body out of the ordinary doorway, an opening was made through the side of the hut sufficiently large for the body to pass and those accompanying it. The relatives and friends followed the remains to the place of the dead, two or three miles distant, upon some hill side. There it remained untouched for five days. The face only of the deceased was exposed. On the return of the family connections to the hut, one of the dogs was killed. During the five days which intervened, the husband forsook the hut altogether, and all other huts, and wandered about from place to place, living in temporary exile from all connection with his former home, or family and friends. And during this time, also, food was carried to the dead body, and also placed outside the hut, on the supposition that she would need it.

On the sixth day, the deceased was visited again by the relatives for the purpose of disposing of what was left of her remains. The crows and beasts of prey had nearly or quite completed the work of destroying every vestige of the body. Thus, in a very short time, nothing remained but here and there a bone mingled indiscriminately with others in the place of the dead.

The company then returned to the hut, and another scene of wailing and mourning ensued. During this last act, the hut was surrounded by the relatives of the deceased; and all at once, at a given signal, the whole company rose up, and pulled the hut down, and removed it to another place. Before it was erected again, however, the second dog was killed, and its blood sprinkled over the newly-selected spot.

With the change in the locality of the hut a new order of things took place. The husband assumed his former relations to the family, and ceremonies were at an end respecting the deceased.

From what could be learned from the natives, they supposed that, in leaving the face of the deceased uncovered, the crows would pick out her eyes, and then she would be unable to find her way back to the hut. The opening made in the side of the hut, through which to carry the corpse, was another superstitious idea. They believed she would not enter the hut again, if she was not carried out by the door. The re moval of the hut to a new place was in accordance with their notions that she would be unable to find it again.

They have a general belief of an existence after death; yet so crude, ill-defined, and dark was this belief, that it stands allied with the grossest forms of paganism and idolatry. The glorious gospel of Christ, "which brings life and immortality to light," finds no place in their hopes for the future, nor does it afford any consolation to them on their pilgrimage to the tomb. They are living, as the apostle said the heathen did in his day, "having no hope, and without God in the world."

MARRIAGES.

They are polygamists. They have as many wives as they see fit to take, or as they can support. They have a custom among them of temporarily exchanging their wives with each other. The evils of polygamy were obvious among the natives, in the jealousy, contention, wrath, and fighting observable between the different wives.

GENERAL APPEARANCE.

From their appearance, we should judge they belonged to the race of Esquimaux. In stature, they are rather below medium height, thick set, strongly built, muscles fully developed, and capable of great endurance; and in complexion, copper color. Their countenances are far from being prepossessing; high cheek bones, flat noses, and large mouths.

A stranger, upon first sight, would be led to infer from their general appearance that they were fierce, cruel, and prepared for any act of barbarity. What they would become, if injured or abused, we had no opportunity of knowing. Nor did we discover in them any unfriendly feelings towards other settlements or tribes, whether near or more remote, or that of late years there had been any contention, or fighting, or war between the different tribes in that region. They had instruments of war, such as bows and arrows, lances, clubs, &c.; but they probably needed them in destroying the savage beasts, and especially bears, that infest the country.

We found them kind and hospitable to us, or otherwise we all must have perished. They treated us, we believe, according to their knowledge and circumstances, with more than ordinary attention.

They exhibited love and sympathy towards the members of their respective families, and were particularly affectionate to their children.

As to their moral character, we could not discover that they had any idea of the one God, the Maker and Upholder of all things and beings, nor of Providence, nor of accountability, nor of moral right and moral wrong.

They believed, however, as all heathen idolaters and pagans do, that there were superior divinities. They seemed to fear evil spirits, if they had no reverence for good ones. They had an idea that somewhere, in some remote mountain in the interior, their god lived, and that the dead would in some way or other go thither; though they never gave us their views particularly upon this subject. They had no idols nor household gods. They paid profound homage to the crow, and regarded it in some sort as sacred. They wore crows' heads as amulets upon their persons. With the exception of the kindness they manifested towards us, and natural affection towards their children and to one another, in their social habits, intellectual ignorance, and moral darkness, they must be classed among the most degraded of the human race.

Their habits of industry correspond with the general features of their character. Thus fishing, hunting, making sleds, training dogs in their teams, running races, occasional traffic with tribes in the interior, &c., constitute the principal routine of their employments and amusements. Their manner of life presented no inducement for them to labor beyond their present necessities. By the way, the females had a large part of the necessary work to perform.

PROPERTY.

Their property consisted chiefly in dogs and huts. He who owned the best dog teams, and had possession in huts, was considered the most wealthy man. The head man of the settlement was supposed not only to possess the greatest amount of property, but he excelled in bodily strength. With these qualifications, he commanded the greatest influence, and was acknowledged as the head and leader of the settlement.

We found some among the natives who were considered rich,—rich in dogs and dog teams, &c.,—and others that were poor.

LANGUAGE.

It is quite probable that all, or nearly all, the tribes or settlements on the shores of the Arctic, both on the Asiatic and American sides, have a common language, though differing, as we found, in some words, and also in pronunciation. The language is that of the Esquimaux race.

Those with whom we lived, and other settlements or tribes on the Asiatic coast with whom we have had any acquaintance, from East Cape to the north as far as our wreck, have no written language. We could not learn from them that any one had ever attempted to instruct them, or reduce their language to some system, or that any teacher in religion had ever visited them. Without a written language, or books, or teachers, or oral instruction in some form, the certain results must invariably be, that from age to age, they will continue in the same condition of mental ignorance, moral blindness, and physical degradation.

It was the opinion of Mr. Abram Osborn, Jr., who became a proficient in the language of the natives, and could converse with them with ease and fluency, that it was simple, and he believed could be readily reduced to some systematic form. The method he resorted to, in order to acquire the language, was simply this: when he heard a native word, he would write it, according to its sound, upon a piece of copper, and place opposite to it its English definition. He made inquiries of the natives as to the meaning of their words, and what they called certain things. In this manner he became very familiar with all the terms and phrases which they used.

The following are a few specimens of native language. English words are placed in the first column, and the Esquimaux in the second.

ENGLISH.	ESQUIMAUX.
Dog,	Attat.
Sled,	Woncoose.
Deer,	Korong.
Fox,	Tricokadlekin.
Legs,	Mingara.
Feet,	Partakou.
Fingers,	Riddlegus.
Arms,	Mingukou.
Hands,	Mungit.
Head,	Eloout.
Hair,	Kidweed.
Nose,	Yacka.
Sick,	Atke.
Death,	Youedlin.
Striking, to kill,	Kittegerayouedlin.
Wrestling,	Mupperrudle.
Dancing and frolicking,	Katepangarrakim.
Soup,	Opanga.
Hat,	Yarang.
Spear,	Pocgan.
Arrow,	Kekimbo.
Whale,	Draow.
Boat,	Atuat.
Ship,	Laloutoutline.
Snow,	Addledadle.
Ice,	Retinute.
Water,	Memut.
Skin,	Naglegin.
Walrus,	Redica.
Woman,	Youan.

Seal,	Mamut.
Hill,	Youket.
Mountain,	Nutamut.
Seamen,	Raumkidlins.
Father,	Etletuen.
Mother,	Etlita.
Son,	Youakek.
Ocean,	Numaumkimmemut.
Land,	Nuteskin.
Sea shore, beach,	Nutanute.
Jacket,	Eran.
Cap,	Kile.
Shoes,	Pomeat.
"O dear me,"	Hokeenonkanum.
Mad,	Anguenipo.
Trousers,	Konitre.

NUMBERS.

One,	Ennan.
Two,	Gera.
Three,	Giro.
Four,	Gerack.
Five,	Miltingum.
Six,	Ennan Miltingum.
Seven,	Gera Miltingum.
Eight,	Amgrokim.
Nine,	Conizinkin.
Ten,	Mingitkim.
Twelve,	Mingitkim Gera Parole.
Thirteen,	Mingitkim Giro Parole.
Fourteen,	Mingitkim Gerack Parole.
Fifteen,	Kiddegitten.
Sixteen,	Kiddegitten Ennan Parole.
Seventeen,	Kiddegitten Gera Parole.
Eighteen,	Kiddegitten Giro Parole.

Nineteen,	Kiddegitten Gerack Parole.
Twenty,	Kalekin.
Thirty,	Kalekin Mingitkim Parole.
Forty,	Gerack Kalekim.
Fifty,	Miltingum Mingitkim Parole.
Sixty,	Gera Kalekim Mingitkim Parole.
Seventy,	Giro Kalekim Mingitkim Parole.
Eighty,	Gerack Kalekim.
Ninety,	Gerack Kalekim Mingitkim Parole.
One Hundred,	Miltingum Kalekin.

ICEBERGS.

Some of the most remarkable phenomena seen in the Northern Ocean, and the manner of their formation, are icebergs. They are greatly feared by seamen, and a contact with them would be equivalent to striking a rock. They are formed far up in the polar region during the intense and protracted cold of winter; and in the change of the season in summer, though ice is always accumulating in high latitudes, they drift with the currents into lower latitudes, where they melt, and finally disappear. They are of varied dimensions, indicating by these facts somewhat the sources whence they come, and wearing every conceivable exterior form.

They are formed by the falling of snow over steep and high cliffs on the borders of the sea; "little by little the incrustations on the shore and cliffs increase to the size of mountains, and then, being torn away from their fastenings, either by the winds, or by their own weight, or by the action of the sea beating against their bases or undermining them, are swept into the ocean, where they continue to accumulate by the falling of snow and frozen water, and finally resemble great islands."

Large masses of ice, which take the form of bergs, are formed along the rocky-bound coast of the Arctic.

On the fall of the tide, after the ocean has been frozen over, the localities of the rocks and ledges are clearly observable. When the tide rises, the superincumbent mass is lifted up, and a new layer is formed underneath. This process goes on with the rise and fall of the tides and the accumulation of ice, until vast ridges, broken and dislocated, assuming every variety of appearance, are thus pressed up to a great height.

We observed the gradual rise of one of these immense piles of ice not far from our winter quarters. It appeared to be more than twenty-five feet above the ordinary ice around it.

The cliffs upon whose sides we have seen icebergs form rise to the enormous height of two to four hundred feet. And the shore was so bold, and the depth of water so great at their bases, that a ship would probably strike her yards against their precipitous sides before she would ground. A vessel, therefore, being dashed against those adamantine walls in a gale of wind, would instantly fly to pieces, and not a seaman would be saved.

"The distance to which icebergs float from the polar regions on the opposite sides of the line is, as may be supposed, very different. Their extreme limit in the northern hemisphere is judged to be about lat. 40°, though they are occasionally seen in lat. 42° N., near the termination of the great bank of Newfoundland, and at the Azores, lat. 42° N., to which they have sometimes drifted from Baffin's Bay.

"But in the other hemisphere, they have been seen, within the last few years, at different points off the Cape of Good Hope, between lat. 36° and 39°. One of these was two miles in circumference and one hundred and fifty feet high, appearing like chalk when the sun was obscured, and having the lustre of refined sugar when the sun was shining upon it. Others rose from two hundred and fifty to three hundred feet above the level of the water, and were therefore of great volume below; since it is ascertained by experiments on the buoyancy of ice floating in sea water, that for every cubic foot seen above, there must at least be eight cubic feet below water."

Captain Sir John Ross saw several icebergs in Baffin's Bay aground in water fifteen hundred feet deep! Many of them are driven down into Hudson's Bay, and, accumulating there, diffuse excessive cold over the entire continent; so that Captain Franklin reports that, at the mouth of Haye's River, which lies in the same latitude as the north of Prussia or the south of Scotland, ice is found every where, in digging wells, in summer, at the depth of four feet. "It is a well-known fact that, every four or five years, a large number of icebergs floating from Greenland double Cape Langaness, and are stranded on the west coast of Iceland. The inhabitants are then aware that their crops of hay will fail in consequence of fogs, which are generated almost incessantly; and the dearth of food is not confined to the land, for the temperature of the water is so changed that the fish entirely desert the coast."

As to the relative thickness of common field ice where it remained unbroken through the winter, we found it varied from ten to twenty-five feet in thickness. We had an opportunity of judging, from the fact that we examined several openings which the natives had made in the ice off East Cape for the purpose of taking seal.

The sudden disappearance of large and extended tracts of ice in the northern seas, in addition to its being carried away by the force of currents towards the south, is attributed by many to its sinking. How ice should sink, when its specific gravity is lighter than water, is a question for the speculative to discuss — unless there be some other preponderating element mingled with it, such as fragments of rock, sand, or gravel.

Whalemen have frequently affirmed that they have not only been surrounded by fields or large tracts of ice at night, but in the morning it had wholly disappeared from the surface of the water. Therefore many have arrived at the conclusion, that in certain states of the ice, in the process of breaking up and thawing, it actually sinks below the surface of the water, if not to the bottom.

There was another phenomenon which we observed. During the coldest season of the year, and in certain states of the atmosphere, the air deposits its moisture in the form of frozen fog. It has the appearance of a fine gossamer netting or icicles, and these are dispersed through the atmosphere, and so extremely minute that they seem to pierce and excoriate the skin; and, especially when the wind blew, it was impossible to face this storm of icy vapor. We have seen a deposition of this frost from four to six inches during the space of twelve hours.

A CASE OF POISONING.

We observed that the natives ate all parts of the bear except the liver. Experience had probably taught them that it was not proper to eat, or, it may be, they had seen the fatal effects of eating it among themselves.

A bear, during the early part of winter, was brought into the settlement, which the natives had killed. Some of our company concluded to make a mess out of the liver, and invite others to partake of the dainty. It was eaten, and the consequences were nearly fatal to all of us who partook of it. It produced distress in our stomachs and diarrhœa.

We find the following in Dr. Kane's "Arctic Explorations:" "When I was out in the Advance, with Captain De Haven, I satisfied myself that it was a vulgar prejudice to regard the liver of the bear as poisonous. I ate of it freely myself, and succeeded in making it a favorite dish with the mess. But I find to my cost that it may be more savory than safe. The cub's liver was my supper last night; and to-day I have the symptoms of poison in full measure — vertigo, diarrhœa, and their concomitants."

A SHIP AFTER A GALE.

CHAPTER VIII.

Provisions of the Natives getting low. — New Calamity threatened. — Health and Strength failing. — Necessity of seeking other Quarters. — The only alternative. — Report of a Wreck. — Parties leave. — Dreadful Traveling and Exposures. — Report by the Natives that our Men were frozen to Death. — An Instance of Treachery. — The Captain and his Party leave. — The Weather. — Traveling. — Thoughts of Home. — Preservation. — One of the Party unable to walk. — Left behind. — Found by the Natives. — The Fate before us. — Division of the Biscuit. — Another fails, sits down, and is frozen to Death. — Reflections. — Captain Norton encourages his only remaining Companion. — Singular Appearance upon the Ice. — Dog Teams. — Part of Mr. Fisher's Company. — Encouragement to our Minds. — Natives unwilling to help us. — The Danger of Riding. — Last Effort. — The Music of Barking Dogs. — Our Manner of Traveling. — Dreadful Condition of our Feet. — Captain Norton falls exhausted. — Native Kindness.

In February, it became apparent to all of us, that the provisions of the natives were getting low; we saw it in our daily fare — diminished in quantity, if not poorer in quality.

A new and unexpected calamity now threatened us. One misfortune after another had followed us since the wreck of the ship; deliverance had failed us when it was just within our reach; disappointment and untold deprivation had taken its place; but now, as if our past trials were only preparatory for another, — one more frightful than any we had contemplated or looked upon, — the question was presented to us in its most distressing form, whether we should remain among the natives, and, from present appearances, *starve* to death, or whether, while any strength remained, we should make one more, and perhaps the last effort to reach some other settlement, where we might get provisions enough to live upon. Our prospects never looked more gloomy than at this time.

We were well assured there were huts down along upon the coast, but how far we could not tell; and therefore it was a most hazardous journey, and altogether uncertain whether any one of us would live to reach them.

We were at this time very much reduced in flesh and strength in consequence of short allowance, and therefore greatly incapacitated to endure the labor and fatigue of traveling through the snow, or to withstand for any considerable season the intense cold which then prevailed. And still further, we were aware there would be no protection for us during the long night we should be out; or, it may be, a number of days and nights we should find no shelter. How many fearful odds were against us! Of this fact we were certain: to remain where we were, we should all perish by degrees with starvation; we came therefore to the conclusion, we could but die if we should venture to travel to the next settlement.

The haggard and emaciated countenances of our companions told but too plainly that a change must take place in our living, or soon we should "go the way of all the earth." If our friends at home could have looked in upon us in this time of our last extremity, they would neither have known us, nor would they have supposed, from our appearance, that we could long survive our misfortunes. It is well that we do not always know either the condition or the sufferings of our fellow-men.

It was about this time, while we were anxiously considering our state — what should be done, in what direction to seek for life — a report reached us by means of the natives, that a ship had been cast away on the coast, from seventy to one hundred miles distant, as near as we could judge.

A single *ham* was brought to the settlement by the natives, which confirmed the truth of the wreck.

This circumstance greatly encouraged us, and determined the first party, consisting of only two, to leave one morning, and to travel in the direction of East Cape. In the afternoon of the same day another party of three left, Mr. Fisher and two others, taking the same course as those did in the morning.

The last party soon came up with the first one, and found the two men nearly exhausted, and overcome by the difficulties of traveling, and by the intenseness of the cold; but by encouragement and hope held out to them, that another day they might find a native settlement, they struggled on through that night. The next day, they pressed on the best they could, making, however, but very slow progress, and seeing but little before them to animate their minds, or to raise up their spirits. They had gone as far as strength, or hope, or the love of life could carry them. They became bewildered, chilled, frost-bitten, and blinded by the flying snow; and as their last resort before they should lie down in death, having given up all prospect of getting any farther, they traveled round and round in a circle; and they were found in this condition when discovered by several natives, who immediately led them to their huts, which were only a mile or two distant.

How these men were kept alive during the time they were exposed to the intense cold of the day, and especially of the cheerless arctic night, seeking the best track they could through an unknown region of valleys, cliffs, ice, snow banks, &c., — how these men were kept alive, is a matter of profound surprise, and certainly one of those instances of special providence in behalf of the needy and suffering sons of men.

Mr. Fisher said, all he had with him to eat by the way, when he left the settlement, was "some burnt coffee in his pocket." The others with him were no better off. It is wonderful that they lived amid so much destitution and exposure. What will not necessity compel men to do! Mr. Fisher, with the rest, asked the natives for something to eat; and he obtained a small piece of frozen whale's blubber. In less than two hours they were brought to the huts, and to their great joy found provisions more abundant. The whole distance they had traveled exceeded twenty miles.

A few days after the departure of Mr. Fisher's company, and the one that preceded his, word was brought to the settlement by some of the traveling natives, that the whole party were frozen to death.

This was sad intelligence indeed, and yet it was what we greatly feared. We, however, had our doubts as to the truth of the report. We had some very strong reasons for suspecting the natives of lying — a habit we perceived identified with another, viz., that of stealing. And yet the report could not fail to produce in all our minds intense solicitude respecting the fate of our companions.

They ventured forth, risking their own lives, in order to find better accommodations for the company. As soon as they should find better quarters, and the prospect of preserving us from starvation, the agreement was, to send us immediate word; and then small companies would follow them from time to time, so as not to discommode a small settlement of only a few huts with our whole number coming into it at once.

The reluctance of the natives in our settlement to assist us in finding new quarters, when they knew their provisions were getting wretchedly low, and when they knew, too, that we had not more than one third of our ordinary fare, and that we were becoming weaker and more emaciated day after day, their reluctance to assist us, or to direct us to the nearest settlement, can be accounted for only on the principle that if we died, they wanted us to die with them; or that they did not desire we should go to any other settlement. What their particular motive was in this respect, we could not satisfactorily ascertain.

During this time of uncertainty concerning the fate of Mr. Fisher and those with him, Captain Norton called the company together, and proposed that another party should go out and look for their companions, and ascertain, if possible, whether they were living or not. Eight or ten days had thus passed away, and nothing was heard from the first party, nor could we learn any thing definitely about them from the natives, though we had reason to believe they knew more about them than we did.

On the supposition, however, that Mr. Fisher and his party had perished by the way, as reported by the natives, and lest those who might follow should meet with the same calamity, and thus party after party be lost in those trackless wastes of the arctic, it was thought advisable, if possible, before any more of us followed, to send word by the traveling natives to all the settlements, both near and more remote, whether five seamen had arrived at any one of them, or whether they had been discovered frozen to death.

Word was sent to Mr. Fisher from Captain Norton by means of pieces of copper written upon with lead, and forwarded by the natives. Mr. Fisher also sent word to Captain Norton in the same manner after his arrival at the settlement; but neither heard from the other, anc therefore both parties were left in painful suspense, and especially those who were left behind.

In this instance we discovered another treachery of the natives towards us, and which we found it impossible to account for, considering their kindness towards us in many other respects.

Since nothing had been heard from Mr. Fisher for many days, the captain stated to his men, that he had made up his mind to leave the settlement, and ascertain if possible the fate of Mr. Fisher, and find better quarters. One thing was certain; he assured them he could not live there; that was out of the question. He was greatly debilitated, had scarcely any thing to eat, and for three days past had not eaten a piece of blubber larger than his three fingers.

Accordingly, on the last day of February, the captain left with a company of three besides himself.

We took our departure at sundown, or late in the afternoon, in order to avoid the effect of sunlight upon our eyes. We learned from the experience of the natives to avoid, if possible, this evil; and hence we took the latter part of the day to commence our perilous journey, and chose darkness rather than light.

Our intentions were to travel until we should find more comfortable quarters, or perish in the attempt. We were sensible that from the severity of the cold, we must travel all the time, night and day; there could be no rest or respite for us, with safety, out of doors. If we should stop for any length of time, or sit down, death would be inevitable.

It was intensely cold when we left — such an air as is felt only in the arctic. The northern lights shone very brightly that night; wind quite high; occasionally the snow flying in dense masses around us; and besides, slumping into the snow from six inches to two feet at almost every step.

Thus we traveled, or rather, as it seemed to us, crawled along during that night, keeping our course by the sea shore as much as we could. We found no well-beaten road, or path, but we had to make one for ourselves; no plain before us, but a rugged and broken surface, both upon the frozen ocean and upon the land; immense piles of snow, wrought into a great variety of forms by the circling winds; indeed the whole scene before us was one of the wildest, grandest, and most terrific, that winter could present to mortal eyes, and such as can be seen only where Winter asserts his undisputed supremacy.

And what a night was that for human beings to be out and exposed, with no covering above us but the bright stars, and the brighter coruscations, as they would flash up from the pole and overspread the northern sky! Then we thought of home,—far distant home,—and friends, and the contrast, the strange contrast between their condition and ours! But words are poor vehicles to convey to the reader the emotions of our minds as we felt the loneliness of our condition, and the dreariness of our prospects on that dreadful night. It will never be effaced from the tablet of our memories, and in our hearts may we ever record, as long as life shall continue, the goodness of God in preserving us, and causing our eyes to behold the light of another day.

About ten o'clock on the following day, one of our number began to exhibit more than ordinary weariness, languor, and stupidity. We found he began to lag behind, and was unable to keep up with us, though we were much exhausted, and only by the greatest possible exertion were we able to keep on our feet. We had not stopped, except for a moment, since we left the settlement.

Tired and overtasked nature, however, could not always endure. We all traveled slowly; but one of our number was really making little or no progress at all. We, who were ahead, would slacken our pace, or return to meet him, assist him, and encourage him to hold out and press on.

This we did many times, but we found it absolutely impossible for him to keep up with us. We had no strength to carry him; this was out of the question; and to attempt to help him along for any considerable time, or to wait for him or stay by him, it was certain we should never get any where, and all die together.

The only alternative, therefore, to which, from necessity, we were brought, was to leave him behind. Sad as was our decision in this instance, yet it was distressingly true that, if we had tarried by the way or sat down, we never should have risen again.

We pressed on for our lives. We soon lost sight of our companion in the distance, either resting or making ineffectual efforts to get along. In leaving him in those wintry wilds, we left him, as we supposed, to die. We saw no chance for his escape.

About eight days from this time, we learned that, a few hours after we left him, he was found by some natives in a perfectly helpless state, and carried by them to a settlement several miles distant, where he was taken care of, and finally joined the company at East Cape.

In regard to those of us who were still able to proceed through the drifted snow, how slight the hope that we should long continue our perilous journey, and how probable that each one of us in turn would lag behind, and finally lie down to rise up no more!

We saw in our companion an example of what our own fate might shortly be. Whatever of heart or hope there was left, the captain encouraged those with him to put forth all their strength and energies, as every thing they held dear on earth — even life itself — was now at stake. If they faltered, death was certain; if they pressed on, there might be some remote chance of safety and of life.

When Captain Norton left the settlement, he took with him as his only supply of food, both for himself and his three companions, three sea biscuits, which he hid away the first of the winter as a last resort, not knowing what necessity the future might bring along with it.

The last and final emergency had now arrived. He therefore took one of the biscuits, and divided it into three parts, retained one for himself, and gave the other two to his companions.

Soon after the division of the biscuit, we found a temporary shelter under the lee of a precipitous and broken line of hills, which extended some distance, and which protected us from the cold and piercing north wind.

Captain Norton never allowed himself to sit

down, because he was convinced, so weak as he was, and nearly worn out, if he should yield to the promptings of his almost exhausted body, and sit down, he would never rise up again; and therefore he continued on his feet, and moving about from place to place. He warned his companions again and again, if they valued life, not to think of finding rest by sitting down, or seeking repose in any manner; if they should, death would shortly ensue. There was "but a step between us and death."

Yet, notwithstanding the entreaties, persuasions, and warnings of Captain Norton, another one was observed to falter and disposed to sit down. Being but a short distance from him, we perceived he made no effort to eat his biscuit, and also exhibited that singular dulness and stupidity which are the silent and stealthy precursors of the sleep of death. He was then sitting down in an easy and natural posture. The captain spoke to him several times; but he gave no answer, nor made any movement of any kind. He went to him immediately, though he was not twelve feet distant, to ascertain the cause, and found what we greatly feared; alas! the poor fellow's eyes were set, his limbs were rigid, the piece of biscuit was still in his hand. He was frozen to death; his mortal life had fled; his spirit had gone to God, who gave it!

In the winding sheet of drifting snow we let him remain. What a scene that was to us! We were struggling for life amid elements of destruction such as but few of our countrymen ever witnessed, and, we trust, never will.

Only two of us were now left to pursue our sad, and in some respects almost hopeless, journey. It seems quite incredible that we should have had any courage to make another effort in struggling forward, after what we had just witnessed, and that, at once, we should not have surrendered ourselves to the fate which appeared to follow and surround us.

The captain said to Cox, his only remaining companion, "The best foot forward now, or we shall be left out here; and to be out one more night, we are gone."

Having traveled two or three miles, as we should judge, from the place where our shipmate died, we discovered something in the distance, from one to two miles, skimming along apparently on the ice, which at first had the appearance of a flock of crows. Cox said to the captain, "The crows have come for us already." But upon further inspection, and the object approaching nearer, it turned out to be four or five dog teams, with three of Mr. Fisher's party and a number of natives, bound back to the settlement to let their companions know that they had found good quarters, and also to bring some of them away with them.

This was cheering news indeed — cheering because Mr. Fisher and his party were alive, cheering because it revived our desponding spirits, and infused new hope into our minds that permanent help was not far off.

Those who accompanied the natives with the dog teams saw at once how nearly exhausted the captain and Cox were, but yet the natives were unwilling to take them to the nearest settlement. And, besides, there would have been as great danger, and perhaps even greater, for us to have ridden on the supposition that the natives had been disposed to carry us, than for us to have walked. We should have been chilled to death, if we had remained still or quiet, in a very short time.

The direction to the nearest settlement on the coast was pointed out to us; and we were put upon the track made by the dog teams, and told that the distance to it was six or eight miles.

The captain told Cox, "We must reach the place before dark; the last effort must now be put forth — the best foot forward." It was now about twelve o'clock, M. We started in the direction of the huts, and traveled on as fast as we could, though at the best very slow. The snow was deep, and hard to travel.

All the mental and physical energy which we possessed was called into requisition to aid us in reaching a resting place before night. It was our last exertion. It was indeed a merciful providence that we happened to meet our friends and the natives, otherwise, beyond a reasonable doubt, we should have perished; but meeting them, however, we received great encouragement to our minds, and, furthermore, knew for a certainty the direction and about the distance of the huts. Without such a stimulus as this, and just the one we needed, — for our lives were suspended upon it, — our last resting place on earth would have been made amid the drifting snows of the arctic.

With severe labor and painful exertion, we finally reached the settlement just at night. Before we saw the huts, which were concealed from our view by banks of snow, we were heralded by the barking of the dogs. We knew, therefore, that we were near the abode of human beings. The sound fell on our ears ten thousand times more sweetly than the music of an Æolian harp.

But we hardly knew how we were carried

through the last part of our journey. Strength was given to us by the great Father of all. It was of the Lord's mercies that we did not yield to final despondency, and utterly despair of ever beholding the countenances of our friends again. Hope and heart were in the ascendant; if they had once fallen, all would have been over with us.

Sometimes we crawled along on our hands and knees; at other times we would fall down, both upon the right hand and upon the left, and it seemed to us that we could not rise; and then, again, we would get up and struggle on. In this manner we traveled miles, and especially the last part of the way. Indeed, our feet had become dreadfully inflamed, and large blisters had formed on the sides of them, which made the labor of walking exceedingly and distressingly difficult.

Captain Norton was so completely overcome and exhausted when he reached the hut, that he fell prostrate upon the floor, unable to advance one step farther, and lay almost senseless.

Not only were our feet inflamed and blistered in the most shocking manner, but our clothes were stiff with frost in consequence of perspiration, by our extraordinary efforts to reach the settlement before night. We were treated with great kindness by the natives; our stiff and frosty clothes were soon exchanged for dry ones. After a season of rest, a good supper was prepared for us, consisting of walrus blubber, deer meat, and "ice cream" made of the fat of the deer mixed with snow.

WHALES RAISED.

CHAPTER IX.

Mr. Fisher's Party a short Distance from this Settlement. — Next Day left for another Settlement. — Our Men arriving in small Companies. — Health improving. — Cross the River. — No Signs of Water. — Settlement. — Ham. — The Wreck of a New Bedford Ship. — When lost, and the Circumstances. — Travel to another Settlement. — The head Man a savage Fellow. — Traveling towards East Cape. — Seaboard Route. — Natives kind. — Begging by the Way. — The Whale Boat. — The Broadside of a Ship. — Ship in the Ice. — Drift Stuff. — Sun's Reflection. — Sore Eyes. — Snow Blindness. — The Blind led with Strings. — Partial Remedy. — East Cape reached. — Cordially received by the Natives.

The night upon which we arrived at the settlement, we learned that Mr. Fisher was only a short distance from us, perhaps four or six miles. Mr. F. heard also by the natives that some of his countrymen had arrived at the settlement below.

The next day we were exceedingly sore and tired, not only indisposed to move, but quite unable so to do. Mr. Fisher, however, having come with several dog teams, accompanied by the natives, in order to carry us to his settlement, persuaded us to go with him, assuring us that he found first rate fare. We accordingly went with him.

This place was called Calushelia, a small settlement upon the seaboard west by north from East River. We remained about twenty days in this settlement, in company with Mr. Fisher and his party. We were now, so far as we could judge, about seventy miles south-east of the place where our ship was wrecked.

Since communication was now fairly open between this settlement and the place where we spent the first part of the winter, and since it was known that intermediate huts were scattered along in this direction, our men began to arrive in small companies of four or five, as they could thus be better accommodated by the way than in larger numbers.

A few weeks only had passed away before there was a very perceptible improvement in the general health of all of us. At this time, two thirds of our entire company had arrived. We thought it advisable, as soon as expedient, to form another party, and proceed still farther towards the south in the direction of East Cape.

Accordingly, the captain, with Fisher, Osborn, Blackadore, Norton, and three others, crossed the river on the ice; the river was just south of us. On both sides of the mouth of this river there were native huts. Where the river discharges its waters into the sea or ocean, it is quite wide, having the appearance of a capacious bay. The river flows towards the north.

At this time, which was in March, we could discover no signs of water either in the river or in the ocean. Both were strongly bound in chains of almost perennial ice.

Having passed over this river, we found a temporary shelter and cordial reception in another settlement. Here we remained a number of days, in consequence of a heavy fall of snow and a severe gale of wind.

It being now towards the middle of March, we could plainly perceive a change in the atmosphere. It is true, we were farther south, which made some difference in the temperature; but the air had lost much of that sharp and piercing sensation which we felt in the winter, and which is experienced, we believe, only in this part of the earth's surface.

Much to our surprise and pleasure, during our abode in this settlement, we were served with ham — a new article of food indeed to us, though we had not a great deal of it, still a most agreeable exchange, if only for one meal, in the place of whale and walrus blubber.

We ascertained that these hams were taken from the wreck of the ship Bramin, of New Bedford, by the natives. This ship, as we afterwards learned, came into collision with another ship off the mouth of East River, during the same gale in which the Citizen was lost. It appeared she was abandoned by her officers and crew, who effected their escape on board of the accompanying ship.

In the concussion which took place her foremast was carried away and otherwise seriously damaged; besides, being near to land, and on a lee shore, it was impossible to save her.

From the position in which we found a portion of her remains, it seemed that, after she was abandoned, she must have beat over a ledge of rocks that stretches across the mouth of tne river, and by the force of the gale driven up the river to the distance of nearly ten miles.

We visited the wreck with the natives, who directed us to the spot. We saw a part of her quarter deck, with the ice piled up around it. We saw, also, upon the shore, close by, some of her timbers and broken casks partly covered up with huge masses and blocks of ice.

It was doubtless the report of this wreck which reached us in our winter quarters. But how far it was east of us, or the circumstances attending the wreck, how many were saved or lost, or whether all were lost, we obtained no satisfactory information from the natives at that time.

Nor did those natives who went with us to the wreck know any thing about the fate of the crew. As they had never seen any of them, nor heard of their being in any of the settlements near by, we naturally inferred that all on board were lost, or that they were immediately taken off of the wreck or from the shore by some accompanying ship.

The night before we left this settlement, and where we were well used, another party of our men arrived. We passed on to another collection of huts, about fifteen in number.

The head man of this settlement, and in whose hut we happened to stop, was one of the most crabbed, savage-like fellows with whom we had met in all our past acquaintance with native life. He appeared to take real delight and satisfaction in degrading and mortifying us all he could. He would cut the meat or blubber, whatever it might be, into small pieces, and reach them to us on the end of a stick, for us to take them, or bite them off as a dog. Indeed, we were treated by him in the same line of courtesy as he treated his dogs.

We quietly submitted to all manner of such ill behavior on his part, simply for the sake of peace and safety. We were completely in his power, and he could use us as he saw fit; and the least we said about it the better. This head man was an exception to all whom we saw among the natives for real ugliness. He was a regular savage. We were glad to be off.

We shortly left this settlement, and passed on towards East Cape, following the direction of the sea coast, which from the river is nearly, as laid down in the chart, in the form of a half circle. We observed that the huts and settlements increased as we came farther south; and sometimes, in course of a day or two, we would pass through several small settlements.

When we became wearied and exhausted by traveling, though it was difficult to make very rapid progress in the snow, or when we were hungry, we would stop, rest ourselves, get some blubber to eat, and then travel on again.

Generally we found the natives ready and willing to help us with what they had. We had nothing to give them in return. We were a company of beggars. They saw our destitution and poverty, and therefore their kindness to us must be attributed to the dictates of human sympathy or pity, which in some way or other shows itself in the most barbarous and uncivilized forms of society.

In our journey upon the coast, we discovered a new whale boat, which the natives had probably drawn out of the reach of the water and ice.

We saw, also, the broadside of a ship in the ice near the shore, supposed to be lost the season before. Another ship was reported to have been seen by some of our party in the ice, some distance from the land, with her masts still standing. There were tracks in that direction in the snow upon the ice, which showed that the natives had been to her with their dog teams.

As we passed along, we saw considerable drift stuff, such as wood, broken casks, &c. We continued on in our course on the coast mostly, finding huts from time to time, in which we obtained provisions for our present necessities, until we came within thirty or forty miles of East Cape, or about half the distance between the river and the cape.

As the spring advanced, the sun was constantly attaining a higher altitude — not only imparting some additional heat, but its rays were powerfully reflected from one dense, unbroken surface of ice and snow, which every where met the eye of the beholder.

A new misfortune now assailed us in the form of sore eyes, or snow blindness, which caused intense pain in them, besides being much swollen. All light, especially bright light, became exceedingly distressing to us. We therefore were compelled to suspend our traveling in a great measure during the middle of the day, and took the morning and afternoon, and even the night time, as more agreeable to our diseased eyes. Our eyes were in such a sad condition that we could not endure the powerful and brilliant reflection of the sunlight upon snow of sparkling and perfect whiteness. While the eyes of all of us were very sore and much swollen, some of our number were so blind that they could not see any thing for several days.

We were very anxious to complete our journey to East Cape. Those, therefore, who could see, and were more fortunate in this particular than others, led along those who were blind with the aid of strings. One or two would take hold of the string, and another would guide them. Thus we worked along for miles in deep snows, through narrow paths, up hills and down declivities, over broken ice, now and then pitching into some cavity concealed by the snow. In this manner we who could see, though our eyes were highly inflamed, led those who could not, both by the hand and with the aid of strings or walrus cord. It was slow and tedious traveling, it is true; yet every mile we gained in the direction of East Cape we felt was bringing us nearer to deliverance. All were animated with the desire to reach this goal of our highest earthly hopes.

And hence, notwithstanding the many obstructions which impeded our course, still with perseverance and unyielding purpose we pressed on our way.

When our eyes were in their worst state, we were compelled to suspend our travels altogether; and when they were better, then we started again, and again led each other with strings, until sight returned to all.

The remedy resorted to in order to cure our eyes was that prescribed by the natives, and which they invariably employ, with considerable success, in the removal of this disease, to which they are subject. Some of us will carry the scars to our graves. An incision was made in the fleshy part of the nose, between the eyes, by a sharp-pointed knife or some other instrument. The effect of this treatment was, that by letting out a small quantity of blood, it reduced the inflammation in our eyes.

We reached the long looked for and wished for East Cape on the 25th day of March, just six months after we were cast away. We would, therefore, as we review the past, — its scenes of danger, exposure, and suffering amid the intensest cold and death-bearing winds of an arctic winter, — gratefully acknowledge the special watchcare of a benignant Providence, which has protected us until the present hour.

Having arrived at East Cape, we were received by the natives with the most cordial welcome. They had heard before we came that a company of shipwrecked mariners was on their way down the coast; and, still further, the report of the wreck, and the uncommon circumstance of so great a number of men having lived with the natives for so many months, had even extended several hundred miles south of East Cape.

The natives in this settlement expressed great joy in seeing Captain Norton, whom they had known before, and with whom they had traded.

Arrangements were made by the head man of the settlement to provide for all the company as they should come along, in small parties, from time to time.

Here, also, we were provided with some new native clothes, such as coats and pants, moccasons and caps.

CHAPTER X.

East Cape, a Point of Observation. — The greater Part of our Men gathered here. — The *Kanaka*. — Weather softening. — Ice still firm. — Arctic Scenes. — Icequakes. — Migratory Fowl. — A Whale discovered. — Gala Time among the Natives. — The Natives thorough Drinkers. — A drunken "Spree." — Cruise into the Country. — Birds-egging. — Incidents. — Native Manner of killing Fowl. — Amusements of the Natives. — Vegetation. — Face of the Country. — Fish. — Fowl. — The Ochotsk Sea and Country.

We had now reached nearly or quite the end of our journey; at least we had attained one of the great objects of our desires and exertions. It was from this place as a point of observation, that most, if not all, the ships passing into the Arctic Ocean on the breaking up of the ice could be seen; and hence this place was the most suitable locality, from which we could be easily taken off.

While the greater part of our company remained at this place, one party of our number, consisting of six or seven, passed down the coast some considerable distance, perhaps twenty miles or more; and another party of four took up their abode at a less distance. Our purpose, and, indeed, the chief one we had in view, was that, in thus distributing ourselves along on the coast, we should be more likely to see and notify any ship or ships that might by chance be early upon the coast, as to the locality and condition of the Citizen's officers and crew.

At this time, which was the first of April, all of the ship's company that had survived the horrors of an arctic winter were at East Cape, and at places just south of it, except one poor *Kanaka*, who, in consequence of frozen feet, was unable to travel with the rest of us, and was therefore left behind about one hundred miles among the natives.

The weather was now softening very perceptibly, though there appeared to be but little diminution of ice and snow. The ocean was not yet broken up, but presented one immovable body of granite ice.

Those who are familiar with arctic scenes well remember the report of concussions between huge masses and blocks of ice, the hoarse and dismal chafings between contending pieces, and their violent agitation by the action of a heavy swell, or winds, or currents; at such times it seemed as if the fabled giants or gods of mythology were engaged in some fierce and terrible encounter.

But now, while the ocean was frozen over with

an incrustation like one of the strata of the earth's surface, we were frequently startled at the deep and prolonged sounds, or rumblings, falling upon our ears like peals of thunder, or discharges of cannon from this sea of solid ice; and then their varied echoes and reverberations would roll away in the distance, forming a most sublime *finale* to the music of an arctic winter.

These icequakes, as we might properly call them, at the north, may be placed in the same chapter with earthquakes, exhibiting on a scale of astonishing magnitude and inconceivable energy the throes of nature.

Fogs began to prevail, and so dense that we could discover an object only a very short distance from us; and besides, so saturating that they were equivalent to rain.

In the months of October and November, various species of birds and sea fowl, with the exception of the crow, which is a permanent fixture in all climates and regions, migrate to the south. In the months of April, May, and June, they return again in immense numbers, beyond all calculation. The air seemed to be alive with the feathered tribe.

The last part of April and the first of May, the snow began to waste away, and objects which had for months been concealed were now made

It was about this time that an incident occurred which created great delight in the minds of the natives. The thawing away of the snow had revealed to their rejoicing eyes a dead whale, which was found three or four miles distant from the settlement. It was probably driven ashore the season before, and thus preserved in the snow and ice. It was a gala time with these simple-hearted and ignorant people. All that could go — men, women, and children, hastened to the dead whale for the purpose of cutting blubber. It furnished a fresh stock of provisions for them; a new bite, far better, we presume, than the old, which had become not only reduced, but rather stale. Nor did we fail of receiving our supply from this newly-cut blubber.

It was deeply interesting, as well as amusing, to witness the zeal of the natives in cutting up the whale, and sledding home the blubber with their dog teams. This was a valuable prize to them, and the staff of life.

As we remarked before in the former part of this narrative, at the time of our visiting the wreck with the natives, they were very inquisitive to know whether we had any thing to drink which would make them dance and sing, and such like. From their gestures, words, and actions, we knew they meant *rum*. In addition to our previous knowledge of their habits in this particular, our further acquaintance with them, for half a year or more, confirmed us in the opinion that they loved ardent spirits, and whenever and wherever they could get it, they would drink to excess.

We found they were no half-hearted, occasional, genteel drinkers. They had no idea of making a quantity of spirits continue its enlivening and kicking effects through several days and weeks; but they wanted, and they would have, if furnished with the means, one grand "*burst up*," one tremendous "*spree*," and that would end it for the present, until the next supply could be obtained. They went on the principle that many others tolerate, "they could not have too much of a good thing."

Some spirits had been brought to the settlement, obtained probably by way of traffic from other tribes in the interior, on the borders of the Ochotsk Sea. When the "fire water" arrived at the settlement, it happened to be in the night time; and before much, if any, of it was drunk, the head man came to Captain Norton, called him up, and wanted he should "take a little," as a token of his respect for the captain, as was supposed. By morning, many of the natives who had drank to excess were laid away as those who belong to the class of quiet ones; but others were noisy, confident, and brave—full of their gabble—rich—possessing the whole creation, and a little more.

Under these circumstances they endeavored to display their agility and strength, and perform wonderful feats; such, for instance, as climbing the pole in the hut. This the head man attempted to do, who was, as we should judge, "three quarters over," and after repeated efforts succeeded in climbing up the pole six or eight feet.

His wife, being actuated by the same impulse, concluded she would follow her husband, and climb up after him. This she did, and had got up only a few feet, when her husband's strength, under these circumstances more quickly developed than lasting, gave out; and yielding to the simple force of gravity which he could not well resist, came down upon the head and shoulders of his wife; and by his accelerated momentum both were brought to the ground in double compound confusion, to the great merriment of those of us who were looking on and observing the progress of the scene. It was one of the most laughable incidents we ever witnessed in our lives.

Again, the natives would display, in the most

boisterous manner, their skill in harpooning or lancing the whale, or walrus, and thus brandish their weapons with uplifted arms, as if they were about to strike their prey.

One of this class was so stimulated with alcoholic strength and courage, that suiting his action to the word or impulse, he threw his spear with all his might into the broadside of one of the huts, and it passed within a short distance of the captain's head. He at once concluded it was time for him to seek a place of safety a little farther off, out of the reach of such dangerous missiles thrown by irresponsible hands. He was careful, however, as it was necessary for our protection, not to exhibit any signs of fear in their presence. They had a regular, thorough-going drunken time.

Allow us to say, that *excessive*, *moderate*, or *occasional* drinking of alcoholic stimulants from love to the "dear creature," makes fools not only of the poor natives inhabiting the shores of the Arctic, but also fools, and greater ones too, of those dwelling in the more civilized, and even Christianized, portions of the earth.

In the month of May Captain Norton took a short cruise into the interior, about one hundred miles, with the head man and several others, accompanied by their dog teams. The settlement he visited was called *Souchou*. The principal food among the natives there was deer meat. Articles of traffic carried from the coast were whalebone, whale, walrus, and seal's blubber; in return the natives bring to the coast deer meat, tobacco, spirits, &c.

In order to while away our time at East Cape before ships would make their appearance, or the ice break up, we would frequently go bird's-egging.

The cliffs facing the ocean were high and steep, and various kinds of birds would make their nests and lay their eggs in the crevices and holes of the rocks. No one could ascend them unaided from the bottom, nor would any one dare descend them from the top of the cliff.

We would, therefore, lower down one and another from the top by means of ropes, or walrus cord, fastened around his body; with this precaution, he could penetrate into recesses in the cliff, and obtain as many eggs as were wanted, and then those at the top would draw him up again. This exercise furnished a source of amusement, emulation, and personal daring; and the last, though not the least, the eggs thus obtained gave an agreeable variety to our "bill of fare."

One incident occurred which came near having

a sad termination. Mr. Osborn fell down a steep declivity ten feet or more, and in his descent he happened to strike a narrow, shelving piece of rock, "just large enough," he said, "to stand upon," which saved him from instant death. The distance below him was more than one hundred and fifty feet.

The Orkney Islands are a famous resort for fowls, which build their nests in the caverns of perpendicular cliffs; and individuals obtain eggs in those places in the manner before described, by letting down one and another from the top of the cliff by means of ropes.

A writer remarks, "I have heard of an individual, who, either from choice or necessity, was accustomed to go alone on these expeditions; supplying the want of confederates above by firmly planting a stout iron bar in the earth, from which he lowered himself. One day, having found a cavern, he imprudently disengaged the rope from his body, and entered the cave with the end of it in his hand. In the eagerness of collecting eggs, however, he slipped his hold of the rope, which immediately swung out several yards beyond his reach. The poor man was struck with horror; no soul was within hearing, nor was it possible to make his voice heard in such a position; the edge of the cliff so projected that he never could be seen from the top, even if any one were to look for him; death seemed inevitable, and he felt the hopelessness of his situation. He remained many hours in a state bordering on stupefaction; at length he resolved to make one effort, which, if unsuccessful, must be fatal. Having commended himself to God, he rushed to the margin of the cave, and sprang into the air, providentially succeeded in grasping the pendulous rope, and was saved."

June had now come, with no very particular incident in the monotony of hut life, except, perhaps, that of increased earnestness and desire to behold once more the sail of a friendly vessel, and once more to tread her decks. As the time approached when we should realize such an event, hours seemed days, and weeks months. We visited again and again the high eminence on East Cape, where we had a commanding view of the ocean, to see if there was any immediate prospect of the ice breaking up and drifting away.

THE MANNER OF KILLING DUCKS AND OTHER FOWL PRACTICED BY THE NATIVES.

In migrating either to the north or south, ducks and other fowl usually fly in large flocks, and generally very low. In order to kill the greatest number, the natives would station themselves at

LOWERING FOR WHALES.

one end of a valley, near the bottom of it; and, if possible, they would conceal themselves from the ducks behind little hillocks. Or if at the termination of a valley there should be a precipitous descent of several feet, the natives would be sure to take their position in such a locality. The fowl usually flew along in valleys running north and south.

Thus stationed, with neither guns, bows, nor arrows, but with a number of small balls in their hands, connected with each other by strings from twelve to eighteen inches in length, they await the approach of their game. When the fowl are sufficiently near for their purpose, the natives rise up, and throw, with singular force and precision, these balls at the flock of passing ducks. At first, when these balls leave their hands, they are all together and compact; but with increasing distance, they will open and spread themselves to the extent of the strings by which they are tied together, and, by the time they get into the midst of the flock, they are fully extended; and then these balls, meeting with resistance, will twist around the necks, legs, and wings of the fowl, and bring them to the ground.

The ducks are killed by the natives, not by cutting off their heads, or breaking their necks, but by pressing the foot upon them until they cease to breathe.

AMUSEMENTS AMONG THE NATIVES.

Dog races are favorite amusements among the natives. Their chief emulation appeared to be who should possess the best dog team. He who beat in a race of several miles obtained a number of fathoms of walrus cord.

These dog teams would sometimes contain as many as twenty dogs. At the time of a race, in which all the settlement was greatly interested, the head man would make a feast, and the most important article on the occasion would be seals' heads; and for some purpose, which they understood better than we did, they would put coals of fire, or brands, in the mouths of these heads.

There was another kind of amusement in which they at times engaged. A circle was formed from twenty-five to fifty feet in diameter, and in this circle any number who chose might enter, and then commenced the race. He who could run the longest, and thus tire the others out, obtained the prize, whatever it might be.

The head man of the settlement would sometimes get up a *woman's* race, and they would run the distance with the greatest fleetness. She who won in the race secured beads, needles, combs, &c.

Probably the most exciting of all their amusements was that of jumping.

A large walrus skin was prepared with holes made in the border of it, as near as possible for the men to stand side by side, taking hold of the skin, lifting it up a foot and a half from the ground, and drawing it tight. An individual who wished to try his or her skill in jumping would get upon this skin. By his exertion, and that of the men who held up the skin, he would jump to a great height; and as he came down, he would meet the upward motion of the skin, which would cause a sudden rebound to the jumper. This process would continue until the individual who was trying his skill in jumping was brought down upon his knees, and then there would be a great shout and laughter. Another would then take his place upon the skin.

This amusement was shared in mutually, both by men and women. Occasionally we would try our skill in jumping. Those most expert among the natives, and especially among the women, would for a long time maintain their standing upon their feet, notwithstanding the efforts of those who held the skin to get them down upon their knees.

VEGETATION.

Of vegetation in the arctic region we can say but little. In the valleys, and along shore, there was a variety of coarse grass. The entire region being destitute of wood, and almost of every species of shrubbery that could really be called such, the soil consequently was exceedingly barren of vines, plants, or flowers. There were, however, a few flowering plants which made their appearance in the transient summer allotted to them.

The rocks were covered with coarse moss, and wherever the sun melted away the snow from the hill sides, or plains, or valleys, a small vine would start into life; this vine afforded, as has before been observed, the only bread of the natives.

We never learned that the natives north of East Cape ever attempted to plant any seed, or to raise any kind of vegetables.

THE FACE OF THE COUNTRY.

The shores of the arctic are bold and rocky, and bordered with high, frowning cliffs. As far as the eye could extend in an inland direction, snow-capped peaks, and finally lofty mountain ranges, filled the whole field of vision.

FISH.

Besides the whale, which is the sovereign of those seas, there are seals, walruses, sharks, narwhal, cod, salmon, &c.

FOWL.

Geese and ducks are abundant, besides a great variety of other water-fowl.

The country bordering on the Ochotsk Sea, a place visited by hundreds of whalemen, presents a scenery in some respects quite different from that of the arctic.

While the surface of the country is uneven, interspersed with hills, valleys, and mountains, yet it is quite well wooded, especially on the seaboard.

As far north as 60° we have found patches of potatoes, turnips, barley, &c. As soon as the snow leaves the earth, numberless wild flowers of every hue and color, and some of them very odorous, immediately start into life and beauty, and adorn both the valley and hill side. And what is most remarkable in the multitude of flowers which follow the line of retreating frost and snow, we find in nature, as in opposite and antagonistical views and principles, that extremes meet.

Vegetation here in this region thrives with the greatest possible rapidity. It seems sometimes to put on the air even of romance, or fiction. One season we were in the Ochotsk Sea, which was the 15th of June, and then we found the country covered with snow; but in less than ten days from that time, the forests were leaved out, and every thing wore the dress of summer.

On the shores of the sea in different localities, we found growing in great profusion, berries of various sorts, such as whortleberry, cranberry, blackberry, mossberry, &c.

We found in the Ochotsk Sea, besides the whale, salmon, trout, cod, eels, butts, and flounders.

In addition to large sea fowl, which were very numerous, an immense number of little birds swarmed the air, some of them of beautiful plumage, and excelling in melodious notes. Many of them were so tame that they would light upon the ship's rigging and yards, and even descend to the deck to pick up crumbs, or little particles of food.

CHAPTER XI.

The Ocean still frozen over on the 22d of June. — On the 24th the Ice began to break up. — Whales appear. — Walrus follow the Ice. — Daily looking for Ships. — Report of our Wreck five hundred Miles below East Cape. — Method of sending News by the Natives. — Ships notified of our Condition. — How. — The Resolution of Captains Jernegan and Goosman. — Arrival of two Ships off East Cape. — Natives first spy them. — Stir in the Settlement. — Happy Day of Deliverance. — Words feeble to express our Joy. — A fit Occasion for Gratitude and Thanks to God. — Preparations to go on board. — The Welcome of Captain Goosman. — Captain Norton with Captain Jernegan. — Crew collected. — Changed our native for sailor Dress. — Liberality of the Officers and Crews in furnishing Clothes. — A Review of the Past. — The Settlement visited. — Dinner. — Arrival at the Islands. — A Card.

On the 22d of June, every thing, so far as the eye could reach, remained the same upon the ocean as in midwinter; and, to appearance, there was no immediate prospect that the ice would break up for some time to come.

On the 24th, however, only two days afterwards, vast masses of ice had left the ocean, water appeared in every direction, though the shore along the coast was piled up with immense blocks and sheets of ice; and in the distance we could clearly perceive varied elevations of icebergs, differing in dimensions and form, scattered here and there upon the surface of the water. The scene presented to our view was that of an awful wreck or convulsion in nature, while those stupendous fragments exhibited the force and energy which had been displayed.

The next day, the 25th, we saw whales close in to the edge of the ice on shore; they remained in the vicinity several days, and then went south again, or to some other part of the ocean, as we supposed, for their food.

The ice having left this region more suddenly than common, greatly disappointed the usual expectations of the natives in taking a large number of walrus, which are highly prized among them. They serve about the same purpose with them as the reindeer does to the Laplander.

The walrus follow the ice; and they are usually found in great abundance among it, and especially upon the edges of extensive tracts both of floating and field ice.

After the ice had passed away, and the ocean was once more free from the embargo which winter had laid upon it, we anxiously and daily looked for some approaching sail. We knew that arctic whalemen would soon be along, and pressing their way up towards higher latitudes, an open sea would invite them to secure a profitable sea son's work.

We afterwards ascertained that intelligence of our condition had been carried down the coast full five hundred miles below or south of East Cape, and that the ships which first touched upon the coast were made acquainted with the fact of the Citizen's wreck, and that her officers and crew were among the natives.

Captain Newal, of the ship Copia, was the first one who heard of the fate of the Citizen.

The method by which the news of the ship's disaster, and the condition of her crew, was conveyed down the coast, is at once striking and significant.

Tracts, those little messengers of truth, become oftentimes the appointed vehicles both of temporal and spiritual blessings. Tracts were found in several huts of the natives, carried thither, we suppose, by seamen; and with the exception of pieces of copper, they were all that could be written upon, and thus the only reliable means of communication. From the first, therefore, the captain and his officers availed themselves of this instrumentality; and, whenever they found a leaf of a book or a tract, or a piece of copper, if opportunity occurred, they would send it down the coast by the natives, carefully enclosed in a piece of walrus or deer skin, giving some account of the shipwreck, officers, and crew, and where they could be found.

We hoped by this means that the news of our condition would, sooner or later, reach the ear of some navigator early on the coast, and thus bring to pass a more speedy deliverance.

In this, as the sequel will show, we were not disappointed. The natives had no idea of written language; and, believing that something of great importance was marked upon either the paper or copper, or both, they preserved it with the greatest care, and almost with superstitious reverence.

They had an impression that we could converse with our countrymen and absent friends in this way, which was true; but what they supposed was, that we could talk to them at any time by those mysterious marks. And hence they frequently urged us to speak to them, and obtain some assistance from them, or that they might send some aid to us.

Mr. Reoy, one of our company, was the first to get on board of a ship, — the Bartholomew Gosnold, — he being down some two hundred miles below East Cape. He therefore gave immediate information as to the locality of the officers and crew of the Citizen.

About this time there were five ships at or near Indian Point, working their way towards the north through the floating ice. The news of the shipwreck was brought to these ships by the natives, bearing in their hands tracts and pieces of copper, written upon by the captain and his officers, stating the wreck, where the company could be found, and their earnest desire to be taken off. The natives approached the ships, lying off a short distance from the shore, holding up in their hands those mysterious parchments, in order to attract the attention of those on board. The story was soon told. The tracts and pieces of copper at once removed all uncertainty which had for months surrounded the fate of the Citizen and the condition of her officers and crew.

The announcement that so many fellow-seamen were still in the land of the living; that they had survived the rigors of an arctic winter; that they were not far up the coast, — less than a day's sail, — and that they were anxiously and hourly looking and waiting for approaching ships, was ènough to stir the deepest sympathies of every mariner's heart.

With the least possible delay, being impelled not only by a sense of duty, but actuated by the most generous and philanthropic sentiments and emotions, Captain Jernegan, of the ship Niger, and Captain Goosman, of the ship Joseph Hayden, left immediately to secure the unfortunate ones on East Cape, firmly resolving, like true sons of the ocean, "We will have them on board before to-morrow night." This was early in the morning. They were distant from East Cape more than fifty miles. With a favoring wind, and success attending their efforts in getting through the drifting ice, they reached the cape next day, about two o'clock in the morning, only a few miles in the offing, and in sight of the settlement.

The natives were the first to spy the ships, and one immediately rushed in and informed Captain Norton they had come.

Though it was an event which we all had long looked for and earnestly desired, and time indeed had rolled heavily on its wheels in bringing the happy day of deliverance, yet when it was announced to us, we could hardly believe it. Somehow or other, having been so long inured to disappointment, we felt for the moment it was too much and too good news to credit.

The native who informed the captain at this time had several times before told him that ships were coming, but which proved false. He was therefore inclined to give but partial credit to his statement now. The native came again and again to the captain with the same report, and manifested so much earnestness and interest that the captain said to Mr. Osborn, who was near him, "There must be something in this fellow's statement; get up, and see if it is so." Besides, there was increasing stir, loud talk, and running hither and thither in the settlement — all of which convinced the doubtful that the ships were indeed in sight, and that the joyous day of deliverance had surely arrived.

Well, we went out of the settlement to see, and it was too true to doubt any longer; the ships were in sight, and standing in towards the land. By this time every man in our company had been aroused, and was on the lookout; and the natives also seemed to partake of the common joy in anticipation of our deliverance being so near.

How feeble are words to express the emotions of gratitude and joy that thrilled through every mind! If tears of sorrow had been shed in months past over prolonged disappointment and subsequent suffering, — if our spirits had become hardened by repeated misfortunes and deprivations, which no language can depict, — tears now fell, prompted by far different feelings; our hearts were no longer indurate, but dissolved like water; and every countenance gave expressions of joyous and exhilarating hope.

What a fit occasion was this for a most hearty and unanimous recognition from all our company of profound gratitude to God that so many of us had been thus far preserved, and were now indulging in the animating prospect of seeing our native homes, relatives, and friends once more! The God of heaven and earth should in this manner be honored and glorified in the presence of pagans, and thus put to silence their vain and imaginary superstitions.

Indeed, one of our number was so deeply affected and overcome with the sight before him, and prompted by a sense of the deliverance which a merciful Providence was working out for him and his companions, that he fell upon his knees and blessed the Lord that he was permitted "to see once more another ship under sail."

The two ships that were in sight, and approaching land, were the Niger, Captain Jernegan, and the Joseph Hayden, Captain Goosman. Preparations were now made, in the most expeditious manner possible, to go on board. The canoes of the natives were got ready; but before any thing could be done towards carrying us to the ships, they first made a fire in the boats

TWO SHIPS NEAR SHORE.

in order to drive out the Evil One; and then, that we might not pollute their boats, some "ice cream" — deer fat and snow — must be given to them to eat, as they supposed. Thus reasoned the natives.

Soon, however, we were on our way to the ships. Most gladly we turned away our eyes from the shore, and turned them towards our better home on the deep.

The ship which was the nearer of the two to the shore, and on board of which Captain Norton and his officers first went, was the Joseph Hayden. When Captain Norton landed on deck, dressed in native costume, unshorn, and uncouth in appearance, as all were, Captain Goosman asked, as well he might, and as any other one would, "Is this *Norton*, captain of the Citizen?" He replied, "He used to be, and probably was now." Captain Goosman then embraced him in true sailor fashion, and cordially welcomed him and his officers to the hospitalities of his ship. In a few days, nearly all of our number were collected from the different settlements, and divided between the two ships.

Captain Norton, being a fellow-townsman and formerly a schoolmate with Captain Jernegan, felt disposed, from this previous acquaintance, to take up his abode on board of the Niger. Every facility and comfort the ship afforded was most cheerfully offered by Captain Jernegan to Captain Norton and those of his officers on board with him.

We soon exchanged the burdensome and unwieldy deer-skin clothes, which had so long identified us with arctic natives and arctic life, for the lighter and more agreeable dress of the sailor.

In supplying our company with such articles of clothing as we needed, (indeed, we were absolutely destitute, having nothing but what we stood in,) the officers and sailors of the respective ships most generously contributed to relieve our present necessities. They rejoiced in the opportunity of effecting the deliverance of their fellow shipwrecked mariners, and considered it one of the most joyous events in their lives that they had done something towards augmenting the sum of human happiness, and thus becoming the means of kindling anew, in many minds, aspirations and hopes which had well nigh become extinguished.

Thus, after a series of sufferings and painful reminiscences, — the loss of our ship, with five of our number at the time of the wreck, and one frozen to death while traveling, — having experienced the dreadful rigors of a northern winter and life among the natives, amid untold filthiness and degradation, — shut out from the hearing and company of friends and the whole civilized world, — after nine months and eight days, on the 4th of July, 1853, we found ourselves safe and happy on the decks of friendly vessels, with excellent accommodations, and all that fellow-seamen could do to make us comfortable and contented.

Four or five days after, the two ships above mentioned, with some others, put into East Cape. The natives came off in their canoes, as usual, to trade.

Captain Norton, with several masters of ships, went ashore, and visited the settlement where he and his men had lived, and called for dinner at one of the huts. His object was that his friends might have some idea of the manner of cooking, as well as the articles of food, among the natives, and how they prepared dinner.

The sight was enough for Captain Jernegan, who left the hut as soon as possible, while his stomach sought to relieve itself by several involuntary throws!

Captain Norton made some little presents to the natives in consideration of their interest in him and his men during their abode with them. He collected various articles from the ships, —

such as needles, combs, tobacco, pipes, &c., — and distributed them among the boys, girls, fathers, and mothers. They were delighted with these unexpected gifts, and expressed their joy in a great many fantastic ways.

The next morning, a violent blow came on, and the Niger was obliged to take her anchor, and go to sea. Several ships parted their chains. The captain remained on board of the Niger most of the season, when an opportunity occurred for him to take passage in the ship Helen Augusta, Captain Fales, bound to the islands. This he did, and arrived at Honolulu on the 5th of October, 1853.

The officers and crew of the Citizen were distributed among the ships in the fleet as their services were needed.

The report of the disaster of the Citizen, and the rescue of her crew, preceded him, and had already reached the islands before Captain Norton's arrival.

The news was brought by a ship which left the Arctic about the middle of the whale season, and touched at San Francisco, and from thence was sent to the islands and to the Atlantic states, to New Bedford, and Edgartown.

A few weeks after the captain's arrival, eighteen of his former crew had come along in different ships, and were well cared for, as shipwrecked seamen, by the American consul at Honolulu. Not long after, the officers, and all the crew with the exception of two, had arrived at the islands.

It is proper here to state the readiness with which aid was proffered in supplying the necessary wants of the destitute among our number. The shipwreck and nine months in the Arctic had left some of our companions absolutely destitute; and when they arrived at the islands, after more than a year's absence at the north, they had but little, if any thing more than in what they stood, or what they had on.

There were not wanting, however, kind friends, willing minds, and generous hearts at the islands, both among the citizens of the place and officers of ships in port, who cheerfully rendered immediate aid to the needy and destitute.

The following Card was published in *The Polynesian* November 19, 1853: —

"Captain Thomas H. Norton, late of the whaleship Citizen, of New Bedford, wrecked in the Arctic Ocean September 25, 1852, returns his thanks to Captain Goosman, of the ship Joseph Hayden, for relief afforded himself and men, in taking them off East Cape, and providing them with necessaries when they were destitute.

" Captain N. would also return his thanks to Captain Fales, of the Helen Augusta, Captains Jernegan, Tilton, Pierce, and Gardner, for many acts of kindness and relief afforded himself and crew.

THOS. H. NORTON.

Nov. 17, 1853."

Having disposed of some oil and provisions which were stored at Hilo, Captain Norton now turned his face towards home, and engaged a passage in the ship Harriet Hoxie, Captain M. Passages, however, were freely and cordially offered to him in other ships.

Before the sailing of the ship, Captain Stott, of the ship Northern Light, proposed to Captain Norton to take his ship for another season, as he himself did not wish to go in her the third season to the north.

Captain Norton had made up his mind, and felt anxious to return home to his family and friends, and reluctant to remain away any longer, since his life had been spared amid so many scenes of trial and suffering through which he had passed.

Upon further reflection, however, Captain

Norton changed his purpose; and the terms which he named to Captain Stott being accepted, he concluded he would try his success in the Northern Light, and, if possible, retrieve his past misfortune.

CHAPTER XII.

A Whaling Community. — Interest felt for absent Ones. — The first Intelligence from the Whaling Fleet. — California Mail. — Further News from the Islands. — "Missing Ships." — No Report of the Citizen. — No Letters. — Fears as to her Safety. — When last spoken with. — Either lost or frozen up in the Arctic. — Supposed Fate of Officers and Crew. — Distressing Suspense. — Hoping against Hope. — Prayer answered. — The first Intelligence from the Citizen. — Joy in Families. — Captain Norton's Arrival at Home, and subsequently the Arrival of his Officers, belonging to this Place.

In a community like ours, in which the chief and principal occupation of the male portion of it is in the whale fishery, there is scarcely a family, and perhaps not one, but has some near or more remote relative absent at sea. It is, therefore, by no means surprising that more than ordinary interest should be felt and manifested in behalf of those who "do business upon the great waters."

In some towns upon the seaboard, the inhabitants are engaged in other kinds of fishery, such as the cod and mackerel fishery; but from the port of Edgartown not a solitary vessel of this description sails. It is wholly whaling, with but few exceptions in case of those who are engaged in the merchant or freighting service, and they sail from other ports.

As a general thing, the boys and young men contemplate whaling as chiefly worthy of their emulation and pursuit. It is identified with their first impressions; and subsequent years only tend to deepen those impressions, and ripen them into irrepressible desire and relish for the whaling business.

A very large proportion of captains and officers belonging to this place and other parts of the island sail in ships owned principally both in New Bedford and Fairhaven. Indeed, they have contributed in no small degree to the commercial prosperity of those places. It is worthy of remark, however, in passing, just to state that the number of vessels connected with the business of whaling, belonging to this place, has doubled within four years. This shows an active spirit of enterprise which we trust will be largely increased in coming years.

The first partial intelligence from the northern whaling fleet usually arrives at home ports some time in the month of October. The early arrival of a ship at the islands, or at some port on the Pacific, from the whale ground, furnishes this report as to the general success of whalemen about the middle of the season — whether the "catch" has been good or moderate, very good or deficient. A few scattering letters are also brought by this early means of conveyance, which, being deposited in the mail, soon find their way across the isthmus, into the hands of relatives and friends at home.

About this time, solicitude begins to be apparent in the inquiries made respecting absent husbands, sons, relatives, and townsmen, as to the probable results of the "whale season," how the ships have done, and the health and lives of those who are abroad.

Every California mail will, therefore, for months to come, be looked for with increasing interest, because it may be the bearer either of joy or sorrow to many hearts and family circles.

In the month of November, still further intelligence is received from the whaling fleet; previous reports are corrected, and additional ones are given. The first section of the fleet has already arrived at the islands.

In the months of December and January, the mail brings still additional news, and more correct than hitherto. The great majority of the ships that intended to touch at the islands on their return from the north are reported at this time. The ordinary vehicles of public intelligence — newspapers and letters, both from the islands and from the Pacific coast — unite in announcing the grand rendezvous or arrival of northern whalemen.

If, now, there should be ships not included in the late report, and from which no recent letters have been received either by owners or relatives, and those ships not having been spoken with by others, they are specially marked as "missing ships," and serious apprehensions begin to be entertained lest some disaster may have befallen them.

The mail in February or March is supposed to bring from the islands and intermediate ports all the reliable information respecting those ships that have arrived during the last four months. Therefore a ship not reported now must have either gone to some other port, or never left the northern seas, or been wrecked and lost.

This was the case with the ship Citizen. There was no account of her arrival at the islands, agreeably to the intention of Captain Norton on his return from the Arctic; his friends at home, therefore, looked for the report of his arrival, if not among the first, certainly among the last.

Besides, there were neither letters from him or

his officers, none to relatives, none to the owners of the ship. Other families had heard from absent ones, and were made to rejoice; those interested in the fate of the Citizen, however, were filled with sadness and sorrow.

The absence of letters was ominous of something fearful and distressing. Captain Clough spoke with Captain Norton on the 23d of September, in the Arctic Ocean, lat. 68° or 69° N.; and this was the last and only intelligence from the missing ship. This occurred, as it appeared, only two days before the wreck of the Citizen. Not having arrived at the islands, nor reported from any other place, the conclusion to which all came was at once reasonable and just — either that the ship was frozen up in the Arctic, or cast away on the coast, and her officers and crew, if living, among the natives.

Reflections of this sort gave confirmation to the worst of fears, and wrought in the minds of relatives and friends, and the community at large, an alternation of some slight hope, on the one hand, that they might after all be safe, and, on the other, the distressing fear that they had completed their last voyage on earth, or perhaps were lingering out a miserable existence amid the rigors of an arctic winter.

How little there was upon which to build a

favorable hope! How weak and superficial the foundation which engrossing and prevailing apprehensions would not instantly sweep away and scatter to the winds! Indeed, whatever conclusions might have been drawn respecting the fate of the Citizen, her officers, and crew, how small encouragement there was in the whole field of imaginary probabilities in their favor, to relieve the minds of those at home from the constantly pressing weight of corroding anxiety and distressing solicitude respecting them!

Uncertainty and suspense with regard to an important event is one of the most trying states of mind in which an individual can be placed. How true this is when the life of some friend appears to be suspended upon the slightest possible contingency! Now indications seem auspicious and hopeful; then, again, adverse and threatening symptoms dissipate every cherished anticipation.

Instances have been recorded in which those who were shipwrecked and threatened with instant death on every side, while the prospect of deliverance was exceedingly small or absolutely cut off, have even desired the approaching crisis, however decisive it might be, whether of life or death, as far less distressing than the dreadful suspense which for hours, and even days, hung over them.

For years, until all hope has at length been abandoned, the civilized world, and especially the commercial part of it, was in a state of profound suspense respecting the fate of Sir John Franklin and his companions, entombed in the Arctic. How much sympathy there was felt and expressed for the distinguished lady of the explorer, who was unwilling to withhold any reasonable and even extraordinary efforts for his deliverance while the faintest color of encouragement existed in his favor! Wealth was poured out like water; and strong, self-denying, adventurous men started up and volunteered their services to traverse again the inhospitable regions of the north; peradventure they might find some traces of the explorers, whether living or dead. Through scores of months of hope and fear, distracting anxiety and painful apprehensions, she suffered for her husband a thousand times more than the certainty which his death would have caused.

It was precisely this state of mind which excited and agitated many families in Edgartown in relation to the uncertainty which surrounded the fate of the Citizen and those who sailed in her.

There was a remote, and yet very slender clinging to a bare possibility that they might be among the living; but at the same time, as if to extinguish every spark of hope, the imagination could hardly picture or conceive a condition in which they could live in the arctic regions. While "hoping against hope," and even beyond it, because hope is the only preservative against despair, yet it seemed as if forlornness stood ready to mock the fugitive idea that it could be well with them.

Thus more than twelve months rolled their rounds, and no ray of light was shed upon this dark event of Providence.

How many times the relatives met to talk over this common affliction and calamity! to mingle together their sympathies, for adversity binds kindred hearts! to send up united desires to God that deliverance in some way which they knew not, but which Infinite Wisdom only knew, might be wrought out for the husband, and sons, and brothers, if still in being!

If meetings and partings, however, brought no outward, substantial relief, nor removed in the least degree the same appalling uncertainty which enshrouded the future, this great truth they learned in many sleepless nights, and tedious days, weeks, and months — that they should "trust in God," and stay themselves upon his mighty hand.

Not only was private prayer offered to Him whose ear is ever attentive, who knows and records the pleadings of every humble worshipper, who marks the beatings of every burdened heart, but the spirit of supplication was manifested in the house of God, and one general desire pervaded the community that He who can "bring light out of darkness," and sustain when all human helpers fail, would grant a great deliverance, and return the absent ones once more to their families and friends.

Prayer was heard; and tidings of good, of hope, and safety were already being borne over the ocean wave, and hastening homewards upon the wings of the wind.

In October, by an early arrival at San Francisco from the Arctic, it was reported that the ship Citizen had been wrecked the year before, in September; that a number of her crew were lost at the time of the wreck; that the captain, officers, and remaining part of the men had wintered among the natives; that they were now on board of several ships in that ocean, and, at the close of the whaling season, they would be at the islands.

This was the first intelligence from the ship, for more than twelve months, which imparted the least reasonable hope to the friends at home. It was indeed hailed with joyous emotions, and profound gratitude to God. It at once lifted off a ponderous load of anxiety, solicitude, and sorrow from many hearts.

Spring and summer, with their singing birds, radiant suns, balmy air, refreshing showers, verdant landscapes, and placid waters, had come and gone; Nature had put off the freshness and beauty of a renewed creation, and once more dressed in her autumnal robes, yet this single item of news, brought from a distant ocean, was the dawn of a brighter day, and the precursor of higher happiness than all the outward world could furnish! It chased away the sorrows of the mind; it breathed new life into the spirits; it taught the hitherto disconsolate ones that the hand of a delivering God should now be recognized and adored. Public sympathy flowed in the same channel with those who could now rejoice, as it was heretofore expressed with those who wept.

November news from the islands confirmed the report of the preceding month, cleared away every doubt which distrust might venture to create, and reassured the wife that her husband was safe, and parents and members of the respective families that their sons and brothers had survived the untold severities of an arctic winter.

About one year from this time, — November 5, 1854, — Captain Norton had arrived at Lahaina, from a cruise in the Ochotsk, in the ship Northern Light, of Fairhaven. He left that port in December for home; and, after a passage of one hundred and ten days, the ship was anchored in the harbor of New Bedford, on the 10th of April, 1855. A day or two after, he reached his native place, to greet relatives, friends, and townsmen, whose apprehensions for a long time had been that they should see his face no more.

The results of his three and a half years' absence from home are briefly these: the first season in the Arctic, in the Citizen, he obtained twenty-six hundred barrels of oil, which were wholly lost when the ship was cast away; seventy barrels of sperm were left at the islands, which he took on his outward-bound passage — this was saved; nine months and eight days among the natives, and taken off in July; the second season on board of other ships, in the capacity of guest and passenger; the third season in the Northern Light, in the Ochotsk Sea, where he obtained twenty-four hundred barrels of oil.

Thus, in misfortune, there was still a good share of prosperity. Not only was one cargo lost, but it was nearly replaced again by another in the ordinary time for which ships are fitted out.

But the greatest and crowning mercy of all was, in returning once more, with such good health, from so many dangers, exposures, and perils which have attended the present voyage.

Mr. John W. Norton arrived home in October, 1855; Mr. John P. Fisher, and Mr. Abram Osborn, Jr., arrived home in April, 1856; — making the time since they sailed from New Bedford, October, 1851, four years for the first, and four and a half years for the last two.

CHAPTER XIII.

The Ocean. — The Seaman's Home. — Confidence of the Mariner in his Ship. — Shipwreck. — Moral and religious Claims of Seamen. — The Spirit of the Age. — Interest in the Mariners' Meeting. — Seaport Places. — Sudden Intelligence. — Seamen remembered elsewhere. — Ships supplied with Books. — Bible and Tract Societies. — Good seed sown. — Field for Usefulness. — The American Seaman. — Concert of Prayer. — All interested. — The most important Reform for Seamen.

WHATEVER pertains to seamen in their adventures, explorations, privations, and disasters, never fails to be of interest to all classes in the community.

The ocean is a vast and mysterious world in itself; a world of mighty waters, grand, sublime; an image of eternity, a scene of wonders and terrors, which no mortal tongue can adequately describe. Man, with his frail bark, borne on its ever restless and heaving bosom, is but a mere particle on the surface of the boundless expanse.

Those, however, whose "home is on the deep," inured both to its smiles and frowns, are familiar with this mode of life, and thus become daily conversant with its varied phases around them.

With a good ship, firm deck beneath hi feet, well manned, plenty of sea room, the experienced mariner fears but little the rising wind or the surging main.

"A storm at sea" which would appall perhaps the heart of a landsman, and lead him to abandon all hope of safety, and that the noble vessel would be utterly incapable of contending with the frightful odds against her, is, to the seaman, who looks calmly on the same scene, only as an ordinary episode in ocean experience; indeed, in some respects, a gale of wind is far preferable to a calm. With what confidence and energy the navigator gives his orders, and is quickly obeyed; soon the faithful ship is trimmed to meet the storm; and true to her native instinct, former antecedents, and original design, she parts the crested billow, and bounds over the waves as a "thing of life"!

The destruction of a dwelling, either by fire or by a tornado, and the inmates flying from threatened death, is a sad calamity; and the occurrence of such an event enlists the sympathies of all who hear of it. But sadder by far is the wreck of a ship at sea, or when cast away upon some remote or hostile shore.

Alas! how frequently it is true, that with the foundering ship, the breaking up of the sailor's home, his house, his refuge, his all, upon the deep, a number of the crew, and sometimes all on board, find a watery grave!

The sufferings incident, in many cases, to shipwrecked mariners, both upon the sea and upon the land, have furnished the most affecting themes of prose and poetry; and their recital uniformly touches an answering chord in every sensitive heart.

We feel that it is due to all classes of seamen, to whom we are so much indebted as the carriers of the products of all climes upon the world's great highway, and by whom we are provided both with the necessaries, and even luxuries, of life, — it is due to them, that their religious wants especially, should claim a share of our attention and interest.

The time was when this class of our fellow-men were thought but little of, and cared less about, in so far as it concerned their religious welfare; but with the progressive spirit of the age in which we now live, the lover of his country, the philanthropist and Christian, cherish a generous solicitude in their behalf. During the meetings of our religious anniversaries, there is no gathering, perhaps, that awakens more general interest than that pertaining to seamen. This fact, in connection with what is being done in the cause of seamen, both at home and abroad, is sufficient to prove that there is a growing, and, we trust, an increasing desire to promote the religious good of the sons of the ocean.

In seaport places, it would be natural to suppose that both the temporal and spiritual welfare of seamen would occupy a prominent place in the minds of the people generally. This is to some extent true. In such localities, especially, one discovers that the trains of thought, general conversation, domestic arrangements, family anxieties, prospects for years to come, all, or nearly all, are shaped and controlled by the leading idea of "*business in great waters.*"

This presiding spirit, as it may be justly termed, pervades every department of life. We meet it at every turn, and are reminded, wherever we go, that we live in a seafaring community. We find this fact verified in public resorts for trade, in the family circle, in the prayer and conference meeting, in the sanctuary, in the chamber of sickness, in the house of mourning, and we read the memorials of it upon the tombstone in the silent repositories of the dead.

There is another feature to a seaport place, and especially to a whaling community, which it would be proper just to mention, and that is, the suddenness with which sad intelligence from absent friends falls upon the ears of those at home.

Many have had painful experience in these particulars. Wives, parents, and relatives have been as suddenly reminded of the decease of those near and dear to them, as would be the change of noonday into the darkness of midnight.

How many hearts have been made to bleed in anguish! how many earthly prospects, hitherto bright, have suddenly become shaded and overcast at such an announcement! Indeed, they shortly expected to hear that those abroad were in health and prosperity; or soon to embrace them on the homeward arrival of the ship; but alas! some mysterious contingency in providence supervened, and terminated their earthly voyage.

Broad oceans, remote seas, distant islands, and foreign ports are consecrated to the memory of seamen, as their last resting places on earth. Indeed, such localities are of impressive and affecting significance, illustrating at once both the nature of the employment and daring adventures of whalemen. But interest for the sailor may not be wholly confined to seaport places. Nor is it. Wherever intelligence reaches, or the public print finds its way through the various avenues of society, or wherever works pertaining to seamen are scattered abroad, even to the farthest limits of civilization, there the sailor will be remembered, and the recorded experience of his ocean life will be read again and again with thrilling emotions. But this is not all. There are hundreds of young men, from inland country towns, and from every part of the United States, whose home is now on the ocean wave, and exposed to the dangers and perils of the deep. It is, therefore, reasonable to suppose, that many a father's and mother's heart follows in affection, hope, and imagination, the absent son upon the unknown waste of waters, or into distant lands; or the wife, anxiously looking for favorable intelligence, offers daily prayer for the successful and speedy return of her husband. Thus, in these respects, those living in the country share in a mutual sympathy with those on the seaboard.

In those places, especially, where large numbers of seamen usually congregate, Bethel services on the Sabbath are means of securing to them a great amount of moral and religious instruction. Besides, colporteur seamen become an efficient instrumentality in directing many a weather-beaten mariner to the house of God, and to the Saviour of sinners.

When whale ships are about to leave our port for a cruise of two, three, or four years, it is the purpose of the friends of seamen connected with the several religious denominations in seaport towns, to place on board of such vessels copies of the Word of God, moral and religious books, the Family Library, tracts, &c. We believe this is the usual practice in other whaling ports;* but to what extent this arrangement is generally carried out we are unable to say.

We acknowledge with gratitude the repeated donations of Bibles and Testaments both from the American and Foreign Bible Society, and the American Bible Society, for gratuitous distribution among the sons of the ocean. Nor would we forget to mention our obligations to the American Tract Society for thousands and tens of thousands of pages of tracts, generously given to be placed on shipboard or put into the hands of seamen. We believe that the good seed of divine truth, thus sown broadcast from year to year, will not wholly fall on unpropitious soil. We are encouraged and strengthened in this benevolent work by the express and significant promises of inspiration: "Cast thy bread upon the waters, for thou shalt find it after many days;" and, again, "Sow beside all waters."

More, however, should and ought to be done

* See the Report of the New Bedford Port Society for 1857.

for seamen. The benevolent and the religious, if so disposed, may find here an ample field for the exercise of their liberality. While something is being done for seamen with reference to their moral and spiritual improvement, yet, when we take into consideration the scores of thousands that yearly leave seaport places in our country, the thousands that are now traversing seas and oceans both near and remote, and visiting almost every part of the earth's surface, how limited are the means employed in behalf of their religious welfare, that Christ may become the pole star of hope to the wandering and tempest-tossed!

The American seaman, in a certain sense, is our representative abroad; and, wherever the stars and stripes are given to the wind and fly from the mast head, there he leaves the impress of his influence. How important it is, then, as he departs from the land of his birth and from the scenes of his early associations, and goes out upon the ocean to meet its dangers and perils, as he is assailed by temptations, or mingles with foreigners in other ports, how immensely important it is, that he should be a true representative of Christian institutions and principles at home, and bear about in his own bosom, amid the vicissitudes of ocean life, the "witness of the Spirit" as his true and lasting treasure!

It is true, there are religious captains, officers, and seamen; but what we earnestly desire is, that the number may be increased a thousand fold. Under the benign influence of the spirit of religion and the fear of God, neither Sabbath breaking, nor profane language, nor vice, nor disorder, nor cruelty, nor mutiny would find a place on shipboard. "Thus officered, manned, and conducted, does any man who believes there is a God, who rules the winds and waves and the monsters of the deep, doubt the success of such a ship?" By no means.

There is another instrumentality fitted to promote the religious interests of seamen, which we would not fail to mention, and that is, the *concert of prayer.* We are taught to "pray for all men;" therefore seamen may be included in the devout supplications of the people of God — not only that they may be mercifully shielded in the hour of danger, and meet with success in all lawful undertakings, but that spiritual blessings in Christ Jesus may be their enduring portion.

The concert of prayer for seamen is one of the most interesting and profitable meetings held in a seaport place, and which the month brings around.

In such a gathering all are interested. Some of the members of congregations and churches are upon the ocean, and have been gone for months and years; others, perhaps, have just left for long voyages, and others still on their homeward-bound passages. At such a meeting as this, the absent ones are brought vividly before the mind. The bare mention of the words husband, son, brother, endeared friend, finds at once a response in many hearts. It is, therefore, alike the dictate of nature as well as the great law of grace to look to Him "whose way is in the sea, and whose path is in the great waters," that he would be with the mariner in the storm and tempest, and at the appointed time return him to his native port and to the bosom of his family.

Temperance and other reforms have wrought, and still are working, gradual and essential changes and improvements among all classes of seamen; but the most important, and that which stands higher than all others, is, that those who behold the wonders of God in the deep may become the friends and followers of the Saviour.

The following hymn, which, with others of like character, is frequently sung, shows at once the sentiments and spirit of the seamen's concert of prayer.

O, pray for the sailor, now far on the billow;
  O, think of his hardships, temptations, and pain.
His home is the ocean; his hammock his pillow;
  He toils for our pleasure; his loss is our gain.

While we are securely and peacefully sleeping,
  He stands at the helm and his duty performs;
Now walking the deck and his painful watch keeping,
  Or sits at the mast head 'mid perils and storms.

O, pray for the sailor, to banishment driven,
  Enduring privation, oppression, and care, —
Shut out from the gospel, a stranger to Heaven,
  The victim to vice and a prey to despair.

And, while we thus pray for the sons of the ocean,
  A kind, peaceful Home to him must be given;
The Mariners' Bethel allures to devotion;
  The Bible and preacher direct him to Heaven.

Seamen, of all classes, you are remembered by thousands and tens of thousands, throughout the land and world, who are deeply interested in your welfare! Day and night you are thought of and prayed for by those whom you have left behind; and many a desire is breathed out in the presence of Him who alone can save, that you may be protected in your absence, shielded from temptations, and returned again to your friends.

May the "Star of Bethlehem, which alone can

"fix the sinner's wandering eye," guide many a son of the ocean, and lead him to say, —

It was my guide, my light, my all;
  It bade my dark forebodings cease;
And, through the storm and danger's thrall,
  It led me to the port of peace.

Now, safely moored, my perils o'er,
  I'll sing, first in night's diadem,
Forever, and forevermore,
  The Star — the Star of Bethlehem!

PERILOUS SITUATION OF WHALEMEN.

A BRIEF

HISTORY OF WHALING.

WITH SOME OF ITS INTERESTING DETAILS.

CHAPTER I.

Whale Fishery. — Its Origin. — Where first carried on. — By whom. — Whaling in the Northern Ocean by the Dutch and English. — Contentions between them. — The Success of the Dutch. — Its Commencement in New England. — " London Documents." — The first Whale Scene in Nantucket. — Boat Whaling. — The Number of Whales taken in one Day. — The first Spermaceti Whale. — The Interest it excited. — Its supposed Value. — The first Sperm Whale captured. — New Life to the Business. — Whaling in Massachusetts in 1771-75. — Burke's Eulogy on New England Whalers. — Sperm Whaling in Great Britain. — Revived in France. — The American and French Revolutions nearly destroyed the Business. — Loss to Nantucket. — Its Commencement in New Bedford. — Tabular View of the Number of Vessels engaged in Whaling, and Places to which they belong.

" No species of fishery, prosecuted any where on the surface of the ocean, can compare in intensity of interest with the whale fishery. The magnitude of the object of the chase, and the perilous character of the seas which it frequents in all climates and latitudes, are features which prominently distinguish the whale fishery from all similar pursuits, and which invest the details of its history with the strong charm inseparable from pictures and verities of stirring exertion, privation, adventure, daring, and danger." In a word, it is fishery upon a gigantic scale, in which romance and reality are strangely blended.

"The whale fishery is a practice of long standing in the world. It is supposed that the Norwegians began to prosecute this hazardous and arduous enterprise as early as the closing part of the ninth century. From rather vague statements, on this subject, which have come down to us, it would seem that they confined themselves to the capturing of a few whales in their bays and harbors.

"The shores of the Bay of Biscay, where the Normans formed early settlements, became famous through them for the whale fishery there carried on. In the same region, it was first made a regular commercial pursuit; and as the whales visited the bay in large numbers, the traffic was convenient and easy.

"The Biscayans maintained it with great vigor and success in the twelfth, thirteenth, and fourteenth centuries.

"We find from a work of Noel, 'Upon the Antiquity of Whale Fishing,' that, in 1261, a tithe was laid upon the *tongues* of whales imported into Bayonne, they being then a highly esteemed species of food. In 1338, Edward III. relinquished to Peter de Puyanne a duty of six pounds sterling laid on each whale brought into the port of Biarritz, to indemnify him for the extraordinary expenses he had incurred in fitting out a fleet for the service of his majesty.

"The Biscayans, however, soon gave up the whale fishery for the want of fish, which ceased to come southward, no longer leaving the icy seas.

"In process of time, voyages both of the Dutch and English were undertaken to discover a passage through the Northern Ocean to India; and though they entirely failed in their primary object, yet they laid open the remote haunts of the whale, and immediately began to prosecute the enterprise of their capture. Even then, it was said, they employed the Biscayans as their harpooners, and for a considerable part of their crew. The Dutch and English prosecuted the business with varied success, each claiming the ground for whale fishery in the seas around Spitzbergen. Large companies were formed, and many ships were sent to those northern regions, each armed and prepared to maintain his right and supremacy over the seas. Thus one party would obtain a charter from its own government, to the exclusion of the other and all others — at the same time, each claiming the prior right of possession by discovery.

"At length, in 1618, a general engagement took place, in which the English were defeated. Hitherto the two governments had allowed the fishing adventurers and companies to fight out their own battles; but in consequence of this event, it was considered prudent by each party to divide the Spitzbergen bay and seas into fishing stations, where the companies might fish and not trouble each other.

"After this period, the Dutch quickly gained a superiority over their rivals. While the English prosecuted the trade sluggishly and with incompetent means, the Dutch turned their fisheries to great account, and, in 1680, had about two hundred and sixty ships and fourteen thousand seamen employed in them." *

"From the year 1660, or forty years after the landing of our pilgrim fathers on the shores of New England, down to the end of the seventeenth century, there seem to have been various, and, as far as now can be ascertained, nearly simultaneous and independent attempts to prosecute this business by the inhabitants of Cape Cod, those of Nantucket and Martha's Vineyard, and some of the British subjects in the bays around the Bermuda Islands."

The following interesting facts respecting the early history of whaling in this country were obtained from manuscripts in the New York State Library, by R. L. Pease, Esq., of Edgartown. They were copied from the originals in London, by

* Chambers.

Mr. Brodhead, under the authority of the State of New York, and called "London Documents."

Vol. iv. pp. 9-12. In the instructions of the Duke of York to his agent, John Lewen, he is directed to "inquire what number of whales have been killed near ye place within six years last past, and what quantities of whale bone and oyle have been made or brought in there, and how much my share hath amounted to in that time. . . . And you are also to informe yourself how many whales are taken and brought in there, commibus annis. Given May 24th, 1680."

Ibid. p. 71. In his answer, Lewen says "that the number of whales killed is never observed by any person, nor the oil or bone."

Ibid. p. 84. General Andros, on this point, states, December 31, 1681, that "very few whales have been driven on ashore but what have been killed and claymed by the whalers; and, if not proved theirs, then claymed by the Indian natives, or Christians clayming the shores in said Indian's right. And tho I have not been wanting in my endeavors, I never could recover any part thereof for his Royal Highness."

Vol. ii. p. 277. "On ye east of Long Island there were 12 or 13 whales taken before ye end of March, and what since wee heare not; here are some dayly seen in the very harbour, sometimes within Nutt Island. Out of the pinnace, the other week, they struck two, but lost both; the irons broke in one, the other broke the warpe. SAMUEL MAVERICK.

July 5, 1669."

"The first whaling expedition from Nantucket was undertaken by some of the original purchasers of the island, the circumstances of which are handed down to us by tradition, and are as follows: A whale of the kind called the 'scragg' came into the harbor, and continued there three days. This excited the curiosity of the people, and led them to devise measures to prevent his return out of the harbor. They accordingly invented, and caused to be wrought for them, a harpoon, with which they attacked and killed the whale. This first success encouraged them to undertake whaling as a permanent business, whales being at that time numerous in the vicinity of the shores.

Finding, however, that the people of Cape Cod had made greater proficiency in the art of whale catching than themselves, the inhabitants, in 1690, sent thither and employed a man by the name of Ichabod Paddock to instruct them in the best manner of killing whales and extracting their oil.

The pursuit of whales was commenced in boats, and was carried on from year to year until it became a principal branch of business to the islanders. The Indians readily joined the whites in this new enterprise; and the most active among them soon became boat steerers and experienced whalemen, and were capable of conducting any part of the business.

Boat whaling from the shore continued until about the year 1760, when the whales became so scarce that it was wholly laid aside.

The greatest number of whales ever killed and brought to the shore in one day was eleven. In 1726, they were very plenty; forty-six were taken during that year — a greater number than ever was obtained in one year either before or since this date.

It is a remarkable fact that, notwithstanding the people had to learn the business and carry it on under many hazardous circumstances, yet not a single white person was known to be killed or drowned in the pursuit of whales in the course of seventy years preceding 1760. The whales hitherto caught near the shores in boats were of the 'right' species.

The first spermaceti whale known to the inhabitants was found dead and ashore on the west end of the island. It caused great excitement — some demanding a part of the prize under one pretence and some under another, and all were anxious to behold so strange an animal.

The natives claimed the whole because they found it; the

whites, to whom the natives made known the discovery, claimed it by a right comprehended, as they affirmed, in the purchase of the island by the original patent. An officer of the crown made his claim to it, and pretended to seize the fish in the name of his majesty, as being property without any particular owner.

After considerable discussion between the contending parties, it was finally settled that the white inhabitants who first found the whale should share the prize equally among themselves.

The teeth, however, which were considered very valuable, had been extracted by a white man and an Indian before any others had any knowledge of the whale.

All difficulty having been settled, a company was then formed that commenced cutting the whale in pieces convenient for transportation to the try works. The sperm procured from the head was thought to be of great value for medicinal purposes. It was used both as an internal and external application; and such was the credulity of the people that they considered it a certain cure for all diseases; it was sought with avidity, and for a while was esteemed to be worth its weight in silver."

"The first sperm whale taken by the Nantucket whalers was killed by Christopher Hussey. He was cruising near the shore for 'right' whales, and was blown off some distance from the land by a strong northerly wind, when he fell in with a school of that species of whale, and killed one, and brought it home.

"At what date this adventure took place is not fully ascertained, but it is supposed that it was not far from 1712. This event imparted new life to the business, for they immediately began to build vessels, of about forty tons, to whale out in the 'deep,' as it was then called, to distinguish it from 'shore whaling.' They fitted three vessels for six weeks, carried a few hogsheads, sufficient to contain the blubber of one whale, and tried out the oil after they returned home.

"In 1715, there were six vessels engaged in the whaling business, (all sloops, from thirty to forty tons burden each,) and which produced an income of nearly five thousand dollars." *

As the enterprise increased, more capital was invested, larger vessels were built, longer voyages were undertaken, and new localities or grounds for whales were discovered.

Fifty years later, — viz., from 1771 to 1775, — Massachusetts alone employed annually one hundred and eighty-three vessels in the North Atlantic Ocean, and one hundred and twenty-one vessels of larger burden in the South Atlantic Ocean.

"Look at the manner," says Burke, (1774,) "in which the New England people carry on the whale fishery. While we follow them among the tumbling mountains of ice, and behold them penetrating into the deepest frozen recesses of Hudson's Bay and Davis's Straits; while we are looking for them beneath the arctic circle, we hear that they have pierced into the opposite region of polar cold — that they are at the antipodes, and engaged under the frozen serpent of the south. Falkland Island, which seems too remote and too romantic an object for the grasp of national ambition, is but a stage and resting place to their victorious industry. Nor is the equinoctial heat more discouraging to them than the accumulated winter of both the poles. We learn that, while some draw the line and strike the harpoon on the coast of Africa, others run the longitude, and pursue their gigantic game, along the coast of Brazil."

Such was the eloquent commendation given to the energy and perseverance of New England whalers by one of the most distinguished of British statesmen.

"The first attempt to establish the sperm whale fishery from Great Britain was made in 1775. Nine years later, the French undertook to revive the prosecution of this business. The king, Louis XVI., fitted out six ships himself from Dunkirk, and procured his experienced harpooners from Nantucket;

* Macy's History of Nantucket

others emulated the example of that monarch; so that, before the French revolution, that nation had forty ships in the service.

"The revolutionary war of the American colonies, and the wars of the French revolution, nearly destroyed this flourishing branch of marine enterprise in both countries. Just previous to the war, Massachusetts employed in this service three hundred vessels and four thousand seamen, about half of whom were from Nantucket alone. During that war, fifteen vessels belonging to this island were lost at sea, and one hundred and thirty-four were captured by the enemy. The loss of life in prison ships and elsewhere, and the immense loss of property, show that Nantucket paid as dearly in the struggle for liberty as any portion of our country.

"It was not until the year 1792, many years after the commencement of the enterprise in Nantucket, Cape Cod, Martha's Vineyard, and other places on the sound, that the attention of the people of New Bedford was turned towards the whale fishery." *

From this date until the present time, no permanent obstruction, with the exception of the war of 1812–1815, has occurred to impede the gradual and increasing interest given to this enterprise, and which now assumes commanding commercial importance, and develops unrivaled energy in its prosecution.

The number of vessels in this country employed in the whale fishery far exceeds that of all others engaged in the same pursuit.

The following tabular view will present to the reader the number and class of vessels engaged in the whale fishery, belonging to their respective places in the United States, as reported in the "Whaleman's Shipping List and Merchant's Transcript" for October, 1856: —

* Christian Review, vol. xii.

Places.	Ships.	Barks.	Brigs.	Sch'rs.	Total.	Tons.
New Bedford	209*	128*			337	122,000
Dartmouth	4	6			10	2,698
Sippican				3	3	319
Westport..........		17			17	3,989
Wareham	1				1	347
Sandwich		1	1		2	292
Fairhaven.........	36	12		1	49	15,927
Mattapoisett.......	1	10		1	12	3,281
Nantucket.........	32	4	1	2	39	12,860
Edgartown	10	4		3	17	4,986
Holmes's Hole.....	2	1	1	1	5	1,349
Falmouth	2	1			3	1,111
Provincetown	1	4	1	16	22	2,792
Orleans		1	2	1	4	638
Beverly		3			3	616
Salem..............		1			1	323
Lynn		1			1	216
Fall River.........		3			3	814
Warren, R. I.......	5	10			15	5,025
Newport		4			4	1,206
Providence	1				1	298
New London	32	14	5	12	63	19,176
Stonington	3	3			6	1,949
Greenport	3	7			10	2,958
Mystic	4	2			6	1,840
Sag Harbor........	5	9	2	2	18	5,252
Cold Spring	3	2			5	2,129
San Francisco.....	4	1	4	4	13	2,500

The whole number of vessels employed in the whale fishery in this country, as before reported, is . . . 670

Number of ships, ... 358

Number of barks, ... 259

Number of brigs 17

Number of schooners 46

The tonnage may be put down at 220,000.

Value of property, at $100 per ton, $20,000,000.

The number of seamen engaged in this business, allowing 30 for each ship, 24 for a bark, 20 for a brig, and 18 for a schooner, would be more than 20,000.

Importations of sperm and whale oil and whalebone into the United States in 1856 are as follows : —

* Ships reckoned at 400 tons, and barks at 300.

Sperm oil, ...	80,941 bbls.
Right whale oil, ...	197,890 bbls.
Whalebone, ...	. 2,592,700 lbs.

CHAPTER II.

The Whale. — Its Zoölogy. — The largest known Animal. — Sperm Whale. — Right Whale. — Finback. — Bowhead.

The Whale is the general name of an order of animals inhabiting the ocean, arranged in zoölogy under the name of *Cete*, or *Cetacæ*, and belonging to the class *Mammalia* in the Linnæan system. This animal is named whale from roundness, or from rolling.

"While living in part or wholly in the ocean, it differs in many important respects from the fish tribes, and it is these peculiarities which render it a link between the creature of the land and of the sea. While it has the power of locomotion in the water, like other fishes, yet in other particulars it has no affinity with them; it is as much a mammal as the ox, or the elephant, or the horse — having warm blood, breathing air, bringing forth living young, and suckling them with true milk."

The whale is the largest of all known animals. Some remarks upon the whale and its varieties will form the subject of the present chapter.

1. The Sperm Whale. The *Cachalot*, or *Physeter Macrocephalus*. The principal species are the black-headed, with a dorsal fin, and the round-headed, without a fin on the back, and with fistula in the snout. This whale is known at a distance by the peculiarity of his "spoutings" or "blows." He can be easily detected by whalemen, if he happens to be in company

with other species of whales. He blows the water or vapor from his nostrils in a single column, to the height, perhaps, of twelve feet, inclining in a forward direction in an angle of forty-five degrees with the horizon, and visible for several miles. There is also a wonderful regularity as to time in which he "blows" — perhaps once in ten minutes. He remains on the surface of the water from forty-five to sixty minutes, and under water about the same time. Unless the whale is frightened, whalemen make quite correct calculation as to the chances of overtaking him, or meeting him, or when he will rise to the surface after he has "turned flukes."

When the sperm whale is near, he can be easily distinguished by the form of his head, unlike any other variety of whale. Its head is enormous in bulk, being fully more than one third of the whole length of its body; and it ends like an abrupt and steep promontory, and is so hard for several feet from its front that it is quite difficult, if not impossible, for an iron to enter it — as impervious, indeed, to a harpoon as a bale of cotton.

Besides, the sperm whale has a hump on his back, which distinguishes him from others. This hump is farther forward than the fin on the finback whale.

Sperm whales have been captured from seventy to ninety feet in length, and from thirty to forty-five feet in circumference round the largest part of their bodies. It is supposed by whalemen, from their appearance, that they live, or some of them at least, to a great age. One writer on this subject thought that the sperm whale would attain the age of many hundred years, and even to a thousand years. This, however, is mere conjecture, because there are no dates or facts upon which to found a correct opinion.

Some whales have been taken having their teeth worn off level with the gums; and then, again, in other instances, part of their teeth have been broken off, or torn out by some violent effort.

The whole number of teeth in a sperm whale is about forty-two; they are wholly in the lower jaw, which alone is movable, with the exception of a natural movement of the entire head of the fish.

The teeth admirably fit into sockets in the upper jaw. When the whale is in search for his food, or contending with his foes, he drops his lower jaw, if he sees fit, nearly to a right angle with the under part of his body, and then brings his jaws together with incredible energy and quickness.

Sperm whales engage in fearful and dreadful struggles and conflicts with each other. One was captured, a few years since, having his lower jaw, which was more than fifteen feet long, and studded with sharp-pointed teeth, twisted entirely around at a right angle with his body; he was swimming in that manner when he was harpooned. This was an instance of a most desperate encounter. Another whale was captured having a part of his enormous jaw broken entirely off. The front and sides of their heads, as well as their bodies, not unfrequently exhibit deep lines or furrows produced by the teeth of some powerful antagonist.

It is supposed that, as the sperm whale advances in age, his head not only retains its ordinary proportions, and to appearance becomes enlarged, but the truth is, the other parts of his body, especially his extremities, do actually diminish in bulk and circumference.

In some instances, more oil has been taken out of the head of a sperm whale than from the other part of his body.

The principal food of the sperm whale is "squid," a molluscous animal. "This is an animal of so curious an order as to merit a word of special notice. The principal peculiarity of this molluscous tribe is the possession of powerful tentacula or arms, ranged round the mouth, and provided with suckers, which give them the power of adhering to rocks or any other substances with surprising tenacity. Some of this tribe attain to a great size, and, as large as the whale is, will furnish it with no contemptible mouthful. In the gullet of one sperm whale, an arm or tentaculum of a sea-squid was found measuring nearly twenty-seven feet long."

Whalemen frequently discover large masses or junks of squid floating about, probably torn in pieces by whales in their search after food. The flesh of the squid is soft, without bones, and somewhat transparent, like the common sunfish seen on our shores. It is said that squid have been seen as large as an ordinary whale. This food for the sperm whale is found in great abundance in the Pacific seas.

2. THE RIGHT WHALE. The whale having this general cognomen belongs to the species of *Balæna Mysticetus*. There are several varieties included in this species, as we shall hereafter observe, and which are distinguished by whalemen both in regard to some external peculiarity as well as the different localities where they are usually found.

The right whale differs from the sperm in the following particulars: his head is sharper, more pointed; he has no "hump" on his back; the column of water which he throws up when he "blows" is divided like the tines of a fork; and it rises from his breathing holes in a perpendicular direction from eight to twenty feet.

The right whale furnishes the bone (baleen) so much in common use, and called "whalebone." This bone is taken from the mouth and upper jaw of the whale, and is set along laterally, in the most exact order, several inches apart, decreasing in length from the centre of his mouth, or the arch of his palate, and becoming shorter farther back, while towards the lips the bone tapers away into mere bristles, forming a loose hanging fringe or border.

At the bottom of this row of bone, where it penetrates the gum, and from eighteen to thirty inches downward, we find a material that resembles coarse hair, entwining and interlacing the bone, and thus forming a sort of network, and so thick that, when the whale closes his lips to press out the water, the smallest kind of fish are caught in the meshes, and are unable to escape.

Indeed, the edges of the bones, or slabs, as they might be termed, are fringed with this coarse hair, and it extends to their extremities, as may be seen in the rough state when landed from whale ships.

The length of the bones or slabs * vary in a great measure according to the size of the fish, though some varieties of this species have larger and better bone than others. The value of the bone is enhanced, as a general thing, in proportion to its length.

The principal food of the right whale is a very small, red fish, called "brit." Immense shoals of these fish are seen on whale grounds; and the water to a great distance, even for miles, becomes colored with them.

When the whale takes his food, he throws open his lips, or lets them fall, and, swimming with great velocity, he scoops up an infinite number of these small fish and others that accompany them, some of them scarcely larger than half of an ordinary sized pea; he then closes his lips, and pressing out the water from his mouth, every particle of solid matter is securely retained within.

"The mouth of the whale is an organ of very wonderful construction. In a large specimen of the race, it may measure, when fully opened, about sixteen feet long, twelve feet high, and ten feet wide — an apartment, in truth, of very good dimensions. Notwithstanding the enormous bulk of this creature, its throat is so narrow that it would choke upon a morsel fitted for the deglutition of an ox. Its food, therefore, must be, as it really is, in very small particles. Such is the wonderful contrivance of nature, and in which we can discover an instance of remarkable wisdom in the Creator and Provider of his creatures."

The right whale does not fight or contend with his mouth

* Average, eight feet; longest, fourteen feet.

or head, as the sperm whale does; but his means of attack and defence are chiefly in his enormous flukes. He will, however, when struck, "root around," as whalemen say, and not unfrequently in this manner upset a boat.

This kind of whale, and other varieties, distinguished by the baleen or bone, have no regular time for remaining on the surface of the water after they "breach," nor in remaining under water after they "turn flukes."

The length of a large right whale is about eighty feet, and some have yielded their captors two hundred and fifty to three hundred barrels of oil.

Such a whale would perhaps weigh not far from eighty tons. Allowing one ox to weigh twenty-five hundred or three thousand pounds, he would weigh down more than fifty of such animals.

And what a sublime sight it must be — and whalemen have often observed it — to see such a prodigious living mass leaping right into the air, clear, altogether out of the water, so that the horizon can be seen between the fish and the ocean! These stupendous exercises and gambols of such huge creatures are termed "breaching."

Sometimes a whale will turn its head downwards, and, moving its tremendous tail high in the air, will lash the water with violence, raising a cloud of vapor, and sending a loud report to the distance of two or three miles. This is called "lobtailing" by whalemen.

The oil of this species of whale is less valuable than the sperm. The "whalebone," which now has an advanced price in the market far beyond any previous value attached to it, is obtained from the mouth of the whale about in proportion of a thousand pounds to a hundred barrels of oil.

3. The Finback Whale. This is a smooth, slim fish — smaller usually than a right whale. He is found in nearly all latitudes. His head and mouth are of the same construction with those of the right whale. This whale is known by whale-

men, when seen at a suitable distance, by his "blows." The column of vapor rises in a single stream in a vertical or perpendicular direction. This fish is termed *finback* on account of a fin on his back, differing in this particular from all other species of whale. The oil obtained from him is of the same quality as the right whale oil.

4. BOWHEAD WHALE. This whale is smooth all over, having no "bonnet on his head," as whalemen say, and as right whales have. Their heads differ in shape somewhat from other whales, and hence the name *bowhead* given to them. This species of whale, so far as known, has never been found except in the Ochotsk Sea and Arctic Ocean.

The Greenland whale, and also the species called the *great rorqual*, are doubtless included in the name which our whalemen give to the bowhead.

There are several other varieties of the whale tribe, and different names are attached to them, such as the "scragg," the "humpback," &c.; but the foregoing are all the kinds whether of interest or profit to whalemen.

CHAPTER III.

Whale Blubber. — Enemies of the Whale. — Affection of the Whale for its Young. — Instances.

WHALE BLUBBER. The following furnishes a succinct statement of whale blubber: "That structure in which the oil is, denominated blubber, is the true skin of the animal, modified, certainly, for the purpose of holding this fluid oil, but still being the true skin. Upon close examination, it is found to consist of an interlacement of fibres, crossing each other in every direction, as in common skin, but more open in texture,

ENEMIES OF THE WHALE.

to leave room for the oil. Taking as an example that of an individual covered with an external layer of fat, we find we can trace the true skin without any difficulty, leaving a thick layer of cellular membrane loaded with fat, of the same nature as that in the other parts of the body; on the contrary, in the whale, it is altogether impossible to raise any layer of skin distinct from the rest of the blubber, however thick it may be; and, in *flensing* a whale the operator removes this blubber or skin from the muscular parts beneath, merely dividing with his spade the connecting cellular membrane.

"Such a structure as this, being firm and elastic in the highest degree, operates like so much india rubber, possessing a density and power of resistance which increases with the pressure. But this thick coating of fat subserves other important uses. An inhabitant of seas where the cold is most intense, yet warm blooded, and dependent for existence on keeping up the animal heat, the whale is furnished in this thick wrapper with a substance which resists the abstraction of heat from the body as fast as it is generated, and thus is kept comfortably warm in the fiercest polar winters. Again, the oil contained in the cells of the skin, being superficially lighter than water, adds to the buoyancy of the animal, and thus saves much muscular exertion in swimming horizontally and in rising to the surface; the bones, being of a porous or spongy texture, have a similar influence."

ENEMIES OF THE WHALE. "The whales, gigantic as they are, and little disposed to injure creatures less in bulk and power than themselves, find, however, to their cost, in common with nobler creatures, that harmlessness is often no defence against violence. Several species of the voracious sharks make the whale the object of their peculiar attacks; the arctic shark is said, with its serrated teeth, to scoop out hemispherical pieces of flesh from the whale's body as big as a man's head, and to proceed without any mercy until its appetite is satiated.

"Another shark, called the thrasher, which is upwards of

twelve feet long, is said to use its muscular tail, which is nearly half its own length, to inflict terrible slaps on the whale; though one would be apt to imagine that if this whipping were all, the huge creature would be more frightened than hurt."

A sperm whale was killed off the coast of Peru several years since, whose sides were found to be greatly bruised, and portions of the blubber were reduced nearly to a fluid state. Two thrashers probably attacked the whale, one on one side of it, and the other on the other, and beat him in the manner above described. This fact shows that thrashers are not only able to injure the whale, but most likely by repeated attacks even to kill it.

"The sword fish, in the long and bony spear that projects from its snout, seems to be furnished with a weapon which may reasonably alarm even the leviathan of the deep, especially as the *will* to use his sword, if we may believe eye witnesses, is in no wise deficient."

Thus sharks, thrashers, and sword fish, in pursuit of the whale, and meeting him at every turn, and in all directions, must be powerful antagonists, even with the monster of the deep; and it is not at all unlikely but that, in the conflicts with him, they finally conquer and destroy him.

But there is another, and, without doubt, the most powerful and persevering enemy with which the right whale has to contend. This is a fish about sixteen feet long, and called by his appropriate name, "Whale Killer." A company of these fish attacking the whale will almost surely overcome and kill him. Besides, the whale appears to be sensible of the superiority of his enemy.

Though the whale can and does frequently elude and outstrip the velocity of the fastest boats of the whalemen, yet, when attacked by "killers," he seems to lose all power of resistance, and submits, without any apparent effort to escape. The "killers," in their relish to fight the whale, have been known to attack a dead one which whalemen had harpooned, and were towing to the ship. And so furious and determined were they, that notwithstanding they were lanced and cut most dreadfully by the whalemen in order to drive them off, yet they finally succeeded in getting the whale, and carried him to the bottom. Old whalemen say that "killers" will eat no part of the whale but his *tongue*. They attack him by the head, and if possible get into his mouth and eat up his tongue. The "killers" are a remarkably active fish, and endowed with a set of sharp teeth which may well constitute them a powerful adversary even to the whale, and whose particular and personal enemy they appear to be.

THE WHALE'S LOVE AND CARE FOR ITS OFFSPRING. The strong affection of the whale towards its young has been many times witnessed by whalemen; and yet the nature of their occupation is such, that they turn this interesting and affecting feature of its character to a most fatal account. They will try to strike the young one with the harpoon, and if they effect this, are sure of the old one, for they will not leave it.

Mr. Scoresby mentions a case where a young whale was struck beside its dam. She seized it and darted off, but the fatal line was fixed in its body. Regardless of all that could be done to her, she remained beside her dying offspring until she was struck again and again, and finally perished. Sometimes, however, she becomes furious on these occasions, and extremely dangerous.

Another writer gives the following account of a case which he witnessed in the Atlantic. Being out with fishing boats, "we saw," says he, "a whale with her calf playing around the coral rocks; the attention which the dam showed to its young, and the care which she took to warn it of danger, were truly affecting. She led it away from the boats, swam around it, and sometimes she would embrace it with her fins, and roll over with it in the waves. We tried to get the 'vantage ground' by going to seaward of her, and by that means drove her into shoal water among the rocks. Aware of the danger and impending fate of her inexperienced offspring, she swam rapidly around it in decreasing circles, evincing the utmost uneasiness and anxiety; but her parental admonitions were unheeded, and it met its fate. The young one was struck and killed, and a harpoon was fixed in the mother. Roused to reckless fury, she flew upon one of the boats, and made her tail descend with irresistible force upon the very centre of our boat, cutting it in two, and killing two of the men; the survivors took to swimming for their lives in all directions. Her subsequent motions were alarmingly furious; but afterwards, exhausted by the quantity of black blood which she threw up, she drew near to her calf, and died by its side, evidently, in her last moments, more occupied with the preservation of her young than of herself."

CHAPTER IV.

Whale Grounds. — Whaling Seasons, and where Species of Whales are found. — Sperm Whale Grounds. —Right Whale Grounds. — Humpbacks and Bowheads, where found. — Right Whale not crossing the Equator. — Arctic Passage for Whales. — Maury's Opinion of the Haunts of the Whale in the Polar Sea. — Confirmed by Dr. Kane. — Vessels fitted for Whaling. — Several Classes. — Time of Sailing. — Arrival at Home. — Length of Voyages. — Seasons and between Seasons.

WHALE GROUNDS, OR PLACES WHERE WHALES MAY BE TAKEN. The following embrace all or nearly all the prominent localities which are familiar to whalemen as whale grounds.

The Charleston ground, Brazil Banks, Tristan de Cuna Islands, Indian Ocean, Sooloo Sea, New Holland, New Zealand, King Mill's Group, Japan and Japan Sea, Peru Coast, Chili Off Shore ground, California, Kodiak, Ochotsk Sea, and Arctic Ocean.

WHALE SEASONS AND THE PLACES WHERE DIFFERENT SPECIES OF WHALE ARE FOUND. Sperm whales are taken in the North and South Atlantic Oceans in every month of the year. Sperm whales are taken on the coast of Chili from November to April, and on the coast of Peru in every month of the year. In the vicinity of the Gallipagos and King Mill's group, sperm whales are found. On the coast of Japan, they may be taken from April to October. They are also taken off New Zealand and Navigator's Island, from September to May. From November to March, there is good sperm whaling south of Java and Lombock. In June and July, sperm whales may be found off the north-west cape of New Holland. March, April, and May are considered good months for sperm whaling off the Bashee Islands, but ships are obliged to leave this ground after that time, in consequence of typhoons. From March to July, there is good ground for sperm whaling in the Sooloo Sea, to the west of the Serengani Islands. In the same months, sperm whales are found off Cape Rivers and Canda, close in to the land. In the Molucca Passage, there is good sperm whaling the year round; the best months, however, are January, February, and March. The English whalemen have taken, in years past, a large number of sperm whales in the Red Sea. The area over which sperm whales roam may include the immense space of the ocean or oceans included between the parallels of 60° of latitude, on both sides of the equator. "The sperm whale is a warm water fish," and, according to the opinion of Maury, though it "has never been known to double the Cape of Good Hope, he doubles Cape Horn."

Right whale season off Tristan de Cuna is from November to March; and from January to March off Crozetts and Desolation Islands. Sperm whales are seldom seen near these islands. Right and sperm whaling off the south coast of New Holland, from October to March. In August, there is good ground for humpback whaling around the Rosemary Islands. Right whales are taken in the Japan Sea from February to

October, but bowhead whales have never been seen there. Right whales are taken on the Kodiak ground from May to September; and from March, or as early as the sea is free from ice, until November, in the Ochotsk Sea. Right whales are found in the *southern* part of the sea, and *bowheads* are found in the *north* and *western* part of it at the same time. Bowhead whales are found and captured in the Arctic Ocean as soon as the ice breaks up, which is usually in June, until October.

The right whale is a cold water fish. It has been found by the examination of "records kept by different ships for hundreds of thousands of days, that the tropical regions of the ocean are to the *right* whale as a sea of fire, through which he cannot pass, and into which he never enters."

It has also been supposed, that since the right whale does not cross the torrid zone, which to him is as a belt of liquid fire through which he cannot pass, therefore "the right whale of the northern hemisphere is a different animal from that of the southern."

It is, however, a well-established fact, "that the same kind of whale which is found off the shores of Greenland, in Baffin's Bay, etc., is also found in the North Pacific, and about Behring Straits; the inference therefore is, that there must be an opening for the passage of whales from one part of the Arctic Ocean to the other."

The following facts are taken from Maury's recent work on "The Physical Geography of the Sea," and cannot fail of being interesting to whalemen, and indeed to all classes of readers: —

"It is the custom among whalers to have their harpoons marked with date and name of the ship; and Dr. Scoresby, in his work on 'Arctic Voyages,' mentions several instances of whales that have been taken near Behring's Straits side with harpoons in them bearing the stamps of ships that were known to cruise on the Baffin's Bay side of the American continent; and as, in one or two instances, a very short time had elapsed between the date of capture in the Pacific and the date when the fish must have been struck on the Atlantic side, it was argued, therefore, that there was a *north-west* passage by which the whales passed from one side to the other, since the stricken animal could not have had the harpoon in him long enough to admit of a passage around either Cape Horn or the Cape of Good Hope.

"Thus the fact was approximately established that the harpooned whales did not pass around Cape Horn or the Cape of Good Hope, for they were of the class that could not cross the equator. In this way we are furnished with circumstantial proof affording the most irrefragable evidence that there is, at times at least, open water communication through the Arctic Sea from one side of the continent to the other; for it is known that the whales cannot travel under the ice for such a great distance as is that from one side of the continent to the other.

"But this did not prove the existence of an *open* sea there; it only established the existence — the occasional existence, if you please — of a channel through which whales had passed. Therefore we felt bound to introduce other evidence before we could expect the reader to admit our proof, and to believe with us in the existence of an open sea in the Arctic Ocean.

"There is an under current setting from the Atlantic through Davis's Strait into the Arctic Ocean, and there is a surface current setting out. Observations have pointed out the existence of an under current there, for navigators tell us of immense icebergs which they have seen drifting rapidly to the north, and against a strong surface current. These icebergs were high above the water, and their depth below, supposing them to be parallelopipeds, was seven times greater than their height above. No doubt they were drifted by a powerful under current."

Dr. Kane reports an open sea north of the parallel of 82°. To reach it, his party crossed a barrier of ice 80 or 10C miles broad. Before reaching this open water, he found the thermometer to show the extreme temperature of 60° below zero. Passing this ice-bound region by traveling north, he stood on the shores of an iceless sea, extending in an unbroken sheet of water as far as the eye could reach towards the pole. Its waves were dashing on the beach with the swell of a boundless ocean. The tides ebbed and flowed in it, and it is apprehended that the tidal wave from the Atlantic can no more pass under this icy barrier to be propagated in seas beyond, than the vibrations of a musical string can pass with its notes a fret upon which the musician has placed his finger. . . . These tides, therefore, must have been born in that cold sea, having their cradle about the north pole. If these statements and deductions be correct, then we infer that most, if not all, the unexplored regions about the pole are covered with deep water; for, were this unexpected area mostly land or shallow water, it could not give birth to regular tides. Indeed, the existence of these tides, with the immense flow and drift which annually take place from the polar seas into the Atlantic, suggests many conjectures concerning the condition of the unexplored regions.

Whalemen have always been puzzled as to the place of breeding for the right whale. It is a cold water animal; and, following up this train of thought, the question is prompted, Is the nursery for the great whale in this polar sea, which has been so set about and hemmed in with a hedge of ice that man may not trespass there? This providential economy is still further suggestive, prompting us to ask, Whence comes the *food* for the young whales there? Do the teeming waters of the Gulf Stream convey it there also, and in channels so far down in the depths of the sea that no enemy may waylay and spoil it on the long journey? These facts therefore lead us to the opinion that the polar sea may be an exhaustless resource for the supply of whales for other seas, as well as a common rendezvous for them during the intense cold of arctic winters. Dr. Kane found the temperature of this polar sea only 36°!

Vessels that are fitted out for the purpose of whaling,

whether for *sperm* or *right* whaling, and the time for which they are fitted, may be classed as follows: —

1. Small vessels, principally schooners, though barks and brigs are included, cruise in the North and South Atlantic Oceans. They are fitted for six to eighteen months, and even two years. 2. Ships and barks that cruise in the South Atlantic and Indian Oceans are usually fitted for two to three years. 3. Ships and barks that cruise on the Peru coast, or Off Shore ground, are fitted for two to four years. 4. Ochotsk Sea and Arctic Ocean whalers are fitted for two, three, and four years. 5. New Zealand whalers, sperm and right, are fitted for two, three, and four years.

THE TIME WHEN WHALING VESSELS SAIL TO THEIR RESPECTIVE WHALE GROUNDS. Ships and barks fitted for the North Pacific, the Ochotsk Sea, the Kodiak, or the Arctic Ocean, usually leave our ports in the fall of the year, so as to make the passage of the Horn, or Cape of Good Hope, in the southern summer; these ships will arrive at the Sandwich Islands in March or April, remain in port a week or two, recruit, and sail to the north. On their return from the north in October and November, and sometimes as late as December, they usually touch at the islands again, take in a fresh supply of provisions, it may be ship their oil home, and sail to some other whale ground in a more southern latitude, either for sperm or right whaling, or both, and continue this cruise until the season comes around for them to go to the north again. The first is called the "regular season" for whaling, and the second "between seasons."

Ships that have completed their voyages, and intend returning home, when they leave the Ochotsk or Arctic, generally touch at the islands, or some other intermediate port, for recruits, and arrive on our coast some time in the spring months, and even as early as February or March, though not generally. The great majority of the ships sail in the autumn, and the largest arrivals are usually in the spring.

The Length of a Whale Voyage is determined by the Number of Seasons. One season in the Ochotsk or Arctic, including the outward and homeward passages, consumes *one year and a half.* Two seasons at the north, including the passages outward and home, and one "between seasons," require *two and a half years.* Three seasons, including the passages and two "between seasons," will require *three and a half years.*

Sperm whalemen, who are not governed by these seasons and between seasons, as right whalers are, are absent from home three and a half and four years, and sometimes longer. Indeed, the success or ill success of whalemen in obtaining oil determines essentially the length of voyages.

CHAPTER V.

Increased Length of Whaling Voyages. — Capital. — Value of Oils and Bone. — Value of several Classes of Whaling Vessels. — "Lay." — Boat's Crew. — Whaleboats. — Approaching a Whale. — Harpooning. — Whale Warp. — Danger when the Line runs out. — Locomotive Power of the Whale. — Lancing. — Flurry. — Cutting in. — Boiling out. — The "Case and Junk." — The Rapidity with which Oil may be taken.

The voyages of all classes of whalemen are much longer and more tedious now than formerly. Whales are more scarce, more easily frightened; they change their grounds or haunts oftener; and besides, the number of vessels engaged in their capture, in all seas, is largely increased, compared with the number twenty years since, or even later.

More capital is now employed in this enterprise than ever before; and, were it not for the greatly advanced prices of oils and bone beyond what they were a few years ago, — taking into account the scarcity of whales, the long time occupied on a voyage, the augmented expense of fitting out ships, in the

HARPOONING A WHALE.

high prices of provisions and other incidentals, — the enterprise could hardly be sustained a single year; and certainly but a few years. Immense losses would pervade all departments of this wide-spread system of commercial operation.

A few years since, the price of *sperm* oil by the quantity was only *fifty* to *seventy-five* cents per gallon; but now it brings *one dollar* and *forty cents* per gallon by the cargo.

Right whale oil was formerly sold as low as *twenty-five* cents per gallon by the cargo; but now it brings in the market *seventy* and *eighty* cents per gallon by the quantity.

Whalebone, which formerly was sold as low as *six cents* per pound, — and almost a drug at that, — in consequence of the increased demand for it, and the various and *extraordinary* uses to which it is applied, now readily commands *eighty* cents per pound.

Thus a cargo of three thousand barrels of sperm oil, at the present market value of the article, will amount to more than one hundred and thirty thousand dollars. A cargo of three thousand barrels of right whale oil, including the bone, will command in the market, as their value now is, more than ninety thousand dollars; a *ship* of four hundred tons burden, fitted for a whaling voyage, may be estimated to be worth from thirty to sixty thousand dollars; a *bark* of three hundred tons, valued from twenty-five to forty-five thousand dollars; a *brig* of two hundred tons, valued from fifteen to twenty-five thousand dollars; a *schooner*, valued from eight to twelve thousand dollars.

A vessel owned by a number of persons, or a company, is usually divided into halves, quarters, eighths, sixteenths, thirty-seconds, sixty-fourths, &c.

The "lay" for which an individual agrees to go on a whaling voyage, is the proportion of oil, or its equivalent in money, according to the current value of oil, which comes to his share at the termination of the voyage. A short voyage and a full ship will be a profitable enterprise. Since each and all on board know their individual lays, all, therefore, have urgent, personal considerations to secure both for themselves and employers the greatest quantity of oil.

The captain's lay is from one tenth to one eighteenth of all the oil which is obtained; the first officer's, or mate's lay, from one seventeenth to one twenty-fifth; the second officer, from one thirtieth to one fortieth; the third mate, from one fortieth to one fiftieth; the fourth mate, from one fiftieth to one sixtieth; four boat steerers, each about one eightieth; "green hands," or those "before the mast," not far from one hundred and seventy-fifth lay.

Each whale boat, when properly *pointed*, has six men. Some ships man five boats, others four; barks four, brigs three, and schooners two and three.

Each vessel carries nearly double the number of whale boats which it needs. The whale boats, which combine lightness and strength, are always kept hanging over the sides and upon the quarters of the ship, ready furnished for pursuit, so that, on the appearance of a whale being announced from aloft, one or more boats can be despatched in less than a minute.

When a boat approaches the whale sufficiently near to strike, which is sometimes close alongside, and at other times on the top of his back, the boat steerer, who has the forward oar, immediately "peaks" it, and taking his position at the head of the boat, with harpoon in hand, he hurls it with all his energy, and generally with such force and precision, that he buries the fatal iron in the body of the whale, and sometimes he is killed almost instantly.

"The harpoon with which the whale is first struck is a most important weapon, made of the toughest iron, somewhat in the form of an anchor, but brought to an edge and point. Instead of steel being employed, as is commonly supposed, the very softest iron is chosen for this important implement, so that it may be scraped to an edge with a knife. A long staff is affixed to the harpoon by which it is wielded. Connected with the harpoon there is a strong line regularly coiled in the tub; when the whale is struck, and is disposed to dart away or dive down to the depths of the ocean, he carries the iron sticking fast by the barbs, while the coiled line runs out with amazing velocity. From a tub near the stern of the boat, it passes around a loggerhead, and over the seats of the oarsmen, to the bow of the boat, and then a sheeve or pulley is provided, over which it passes to the whale. The friction sometimes is so great in consequence of the rapidity with which the line is carried out by the whale, if by accident it gets out of its place, the bow of the boat is speedily enveloped in smoke, and would burst into a flame provided water was not instantly applied to prevent or allay all friction.

"It is at such a time as this, when by some slight accident the line gets 'foul,' or, by the overturning of the boat, the warp becomes 'tangled' up with the men, many a poor sailor has been carried out of the boat, and carried down into the depths below, and never seen after. Such sad occurrences as these are not wholly unfamiliar with whalemen.

"As soon as the whale is struck, orders are given to 'stern all,' in order to get out of the way of his flukes, or if he is disposed to be frantic and run, to give him the line. Sometimes the lines of several boats are bent on, and more than eight hundred fathoms are run out, and yet the whale would sink the boats were not the line cut. The force that can drag more than three thousand feet of whale warp through the water, including a whale boat, and sometimes more than one, at the rate of *ten, twelve,* and *fourteen* miles per hour, must be tremendous. Such is the locomotive energy of the whale. It is supposed that with equal ease he could swim off with a ship.

"When, however, the whale becomes so exhausted, having been perhaps harpooned by some other boats, that the warp can be hauled in, and the boat or boats approach the whale again, the lancer, who is generally one of the mates of the ship, exchanges places with the boat steerer, and takes his position at the bow of the boat, with a lance ten or twelve feet long; as soon as he comes near enough to reach him, he thrusts the slender and fatal steel into the very vitals of the animal; 'blood mixed with water is discharged from the blow holes, and presently streams of blood alone are ejected, which frequently drench the boats and men, and cover the sea far around. Sometimes the last agony of the victim is marked by convulsive motions with the tail, and violent contortions of his whole body; and, as we have seen, in its dying moments it turns its rage towards the authors of its sufferings. The whale is now in his 'flurry;' he dashes hither and thither, snaps convulsively with his huge jaws, rolls over and over, coiling the line around his body, or leaps completely out of the water. The boats are often upset, broken into fragments, and the men wounded or drowned. The poor animal whirls rapidly around in unconsciousness, in a portion of a circle, rolls over on its side, and is still in death. At other times, after it is lanced, the whale yields up its life quietly, and dies with scarcely a struggle."

Besides harpoons, which are the most important instruments upon which whalemen depend for capturing the whale, the harpoon gun and bomb lance are now used for the same purpose. They are not, however, considered as substitutes for the harpoon, except in cases of emergency, when the whale cannot be approached by a boat, or when he manifests ugliness or ferocity. The harpoon gun, designed to throw a harpoon, is but little used by American ships, though quite generally among English whalers. Nearly all of our whale ships, however, are supplied with the fatal and destructive bomb lance. The gun, into which the lance exactly fits, is heavier, shorter, and its barrel larger than common guns. It is loaded with powder, in the same manner as other guns. The lance is then put into the barrel of the gun, until one end of it comes in contact with the charge of powder; the opposite extremity has three edges, sharp, and tapering to a point. The entire length of the lance is about eighteen inches. The lance is prepared with a hollow tube, extending half or two thirds of the distance through it; and this tube is filled with a combustible material that readily ignites when the gun is fired. When the lance has buried itself in the huge body of the whale, the fire communicates with the explosive part of the filling in the tube, situated about in the centre of the lance, and in a few moments, thirty seconds perhaps, it bursts like a bomb, and destroys the life of the whale. The bomb lance may be fired with effect at a whale, at a distance of about fifty yards or more.

"The huge body is now towed to the ship; a hole is cut into the blubber near the head, into which a strong hook is inserted — a difficult and dangerous operation. A strong tension is then applied to this hook, and by it the blubber is hoisted up, as it is generally cut by the spades in a spiral strip, going round and round the body, the whale being secured alongside of the ship, and somewhat stretched by tackles both at the head and tail. As this strip or band of blubber is pulled off, weighing from one half to two tons, the body of course revolves, until the stripping reaches the 'small,' when it will turn no more.

"The head, which at the commencement of the process was cut off and secured astern, is now hoisted into a perpendicular position, the front of the muzzle opened, and the oil dipped out of the case by a bucket at the end of a pole." A ship has no purchase sufficiently strong to hoist in on deck the head of a large sperm whale. It is so heavy that it would take the masts out of her if attempted, or bring her keel out of water. Besides, it is so bulky that it would more than fill up the entire waist of the ship. The head sometimes contains more than fifty barrels of oil.

After the oil has been dipped out of the "case," the "junk" is then cut into oblong pieces and taken in on deck; the remainder of the head and carcass are then cut adrift. The oil is afterwards extracted from the blubber and junk, being cut into small pieces by the "mincing knife," and exposed to the action of fire in large pots, the skinny portions which remain serving for fuel. It should be observed that it is usual to secure the "junk" before dipping the oil from the "case." The "junk," which is the forward part of the head, contains the purest spermaceti, and therefore more valuable on that account. It is deposited in the front part of the head in a solid mass, about the consistence of lard, and divided occasionally by a narrow layer of "white horse," a substance resembling the cords of animals, only harder. After passing through a "cooler," the oil is conveyed through leathern hose to large stationary casks which constitute the bottom tier in the hold of the ship. When whales are plenty, which is the harvest time with whalemen, they usually stow away one hundred barrels of oil in twenty-four hours. At such times as these, the fires in the "try works" never go out. If whales were abundant, whalemen would fill a ship carrying three thousand barrels in less than two months.

CHAPTER VI.

Outfitting and Infitting. — "Runners." — Remedy. — Articles of Clothing. — Whaling Business. — Promotion. — Whale Killing. — Dangers. — General Success of the Enterprise.

In connection with the enterprise of whaling, a system of *outfitting* and *infitting*, as they are termed in common parlance, has sprung up, become established, and which is now closely identified and associated with it. This system, from its novel and somewhat singular operation, is like the vine, which entwines itself around the huge and gigantic oak, and thus it grows and expands according to the height and dimensions of

CUTTING IN A WHALE.

its support. Such is the outfitting and infitting business in its relations to whaling.

There are many establishments of this sort, in those places where whaling is carried on, whose principal business is to fit out recruits for whale ships. Hundreds, and perhaps thousands, of young men from the country, who have a desire to go to sea, and particularly whaling, naturally direct their steps to seaport places. There are others, also, who compose the floating, shifting, and in many cases the vicious class of young men, such as are found in all our large cities and prominent seaport towns; these, as a last resort, and in keeping with their roving and roaming habits, enlist in the whaling service. Such, too, are generally poor, wanderers it may be from good homes, becoming associated with bad company, and having no particular means of helping themselves in the time of emergency; therefore they are willing to be assisted in any way by others. Indeed, a change to them is a new fortune.

Advertisements or handbills sent abroad from place to place, proclaiming the want of seamen, are the measures usually adopted. besides some others, for collecting the materials which supply, to a considerable extent, the whaling fleet with "green hands." The outfitters take the general charge of these men, pay their board bills and other incidentals while in port, or before going to sea, and thus supply agents of ships in want of seamen. Scores and hundreds are shipped in this manner who never see the vessels in which they are to sail until they go on board for the voyage.

The outfit is supposed to embrace such articles of clothing, as to quality and value, which seamen need for the cruise, whether longer or shorter, according to the time for which they are shipped. There is scarcely one young man, unless he has had some previous information on this point, or is otherwise familiar with the facts, who knows what he most needs in the line of clothing for a voyage of two, three, and four years.

The outfitter, however, is supposed to know just what the

young man needs. He is therefore provided with a sea chest, and in the chest, his stock, or outfit of clothing, is supposed to be placed by the outfitter, according to the amount for which the respective agents of ships wanting men will be responsible, and for which agents will settle with the outfitters after the sailing of their ships. Outfitters are thus limited by agents of ships as to the amount of the bills of clothing charged against each seaman respectively. The amounts of the bill of goods, or outfit, authorized by an agent, and so understood by the outfitter, will average from sixty to one hundred and twenty-five dollars to each seaman, or some of the lower officers, as boat steerers or fourth mates. Besides, all the expenses which the outfitter has been at in procuring men, and while on their hands before the ship sails, are charged in the several bills against the seamen.

After the sailing of the ship, the outfitter presents his bills to the agent, which he has against the men whom he has furnished for the ship, and these bills are immediately settled. Now, the amount of the bills thus paid to the outfitter is charged by the agent of the ship to each seaman, according to his bill of outfit, or which the outfitter has against him. In the transfer of the bills from the outfitter to the agent who settles them, the agent adds twenty per cent. to each seaman's bill; and thus the seaman, by this change, becomes indebted to the owners of the ship in which he sails.

The outfitter, however, must see his men on board of the ship before she sails; if they are not there, or if they have taken "a land tack," which they sometimes do, clothes and all, the outfitter is the chief and only loser in the affair. Special care, therefore, is taken by the outfitter, that the chests of clothing belonging to seamen shall accompany them when they go on board to go to sea.

Again, seamen are furnished for the whaling fleet by another method: an agent, for example, wishes to procure a certain number of whalemen; and for this purpose he sends to an outfitter, who secures the number that is wanted, gives them an outfit, as before noticed, and places them on board a day or two before the ship sails. This course is now usually adopted with reference to ordinary seamen or green hands. Thus we see the operation of the *outfitting* system.

We would respectfully suggest in this connection, that in our opinion, this method of supplying ships with "fresh hands" is one of the most prolific sources of unhappiness, discord, and every evil work, which not unfrequently take place between officers and crews. The very lowest dregs of society in this way are thus placed on shipboard as foremast hands; and among them there will be found those of desperate characters, and prepared for every work of disturbance and crime.

The *infitting* may be stated in the following brief manner. When a ship arrives in port from a whaling voyage, there are individuals ready to go on board before she approaches the wharf, or even casts anchor in the outer harbor, whose object is to supply seamen, or those whom they have formerly outfitted, as soon as they come ashore, with new clothes; or, in other words, to give them a regular infit. These individuals are called by agents, whalemen, and others, "runners," or "sharks," and are connected with the outfitting and infitting establishments. The seamen are soon provided with new suits of clothing from head to foot, which they greatly need after a three or four years' voyage around the Horn. The results of the voyage, however, if any thing shall be due to the returned seamen at the time of settlement with the agent, are held available to the outfitter; he looks to this source wholly, to meet this additional bill of clothing, or infit, which he has against the young whaleman.

If this were all upon which the "sharks" were disposed to lay their hands, it might be construed into a virtue, perhaps, instead of a fault. But could the history of large numbers of returned seamen, both whalemen and others, be only partially opened and spread out before the public eye, as it not unfrequently is, in that history we should find scenes of temptation, dissipation, and vice, in which not only the hard-earned fruits of years of toil, but character likewise, reputation, and happiness, have disappeared before the voracious grasp of those who lie in wait to destroy.

There are, doubtless, honorable and creditable men in the outfitting and infitting business, as well as in other avocations and callings. Such we do not mean. It is not so much the enterprise as it is the disreputable proceedings of those who are bent on securing unrighteous gain, and to whom, in far too many instances, alas! the unsuspecting sailor falls an easy prey. It is persons of this description, called "runners," or "sharks," that are not even allowed on board of some ships when they come into port, and before the crew are discharged. The purpose of their visits is well known, both to the officers and owners, and therefore they are denied the liberty of coming on board.

Seamen, beware! There are shoals, quicksands, and death-pointed rocks upon the land as well as upon the ocean! Be not led astray. Be men, upright, honest. Shun the cup, and all the gilded and winning blandishments that line the pathway to ruin! Husband, with becoming interest and economy, the results of your toil. Remember that virtue, and the fear of God, united with a conscientious discharge of your duty, both upon the sea and upon the land, will be a sure precursor to happiness, usefulness, and success in life. Take this course, and we assure you, as friends to your temporal and religious welfare, you will escape many a snare spread for your feet, into which others, with less circumspection and watchfulness, sadly and fatally fall.

It sometimes happens, that a seaman who has been on a voyage of several years, finds on his return that he has not made enough to pay his outfit and infit, nor money enough in his pocket to get home to his relatives and friends in the country. The voyage, perhaps, had been an unsuccessful one, and he, therefore, with others, suffers a common loss. Worthy young men experience such instances of misfortune as these; having made little or nothing during their absence from home, they are induced, from a sense of mortified pride, perhaps, to remain away years longer, hoping thereby to gain during the next voyage what they failed to secure in the last one. Thus they ship again, and go through nearly the same routine, the second time, as they did the first; with this exception, however, if they have given proof of efficiency and aptitude in whaling, they will be promoted to the position of boat steerers, and even to higher offices.

The writer on one occasion conversed with a young man, on board of one of our outward bound whale ships, respecting his parents, the place of his nativity, how long he had been in the whaling business, when he left home, &c. He informed the writer that he had a widowed mother in an adjoining state; that he returned from sea in June last, and having made nothing, he was therefore unable to go home and see his mother. Soon after his arrival in New Bedford he shipped again, and is now on another cruise of three and a half years. When allusion was made to his mother, and that in some way he ought to have gone and seen her, the tear instantly gathered in his eye, which showed that beneath a weather-beaten exterior there was something in his bosom which quickly responded to the endearing name — *mother*.

The system of outfitting, to which allusion has been made, and which might be carried on with honesty and integrity, yet nevertheless, as all must see, furnishes an opportunity for the unprincipled and avaricious to defraud and grossly cheat the ignorant and unsuspecting. The following are the ways in which it may be done. 1. *In the poor and miserable quality of cloth of which seamen's garments are made.* They have been known to fall to pieces after being worn only a few times, which clearly proved that the material called cloth was just strong enough to be put into the *shape* of clothes, and that was all. It was poor and cheap, and the buyer of the article probably knew it; it being for whalemen, and outfits justified the purchase. 2. *In the loose and imperfect manner in which seamen's garments are put together.* This is not true of all. The price paid for making is the minimum, or starving price; and therefore the garments are made accordingly. There are two losers by this arrangement, and one winner. The maker and buyer are the losers, while the profit passes into the hands of the seller. 3. *In the exorbitant charges which are sometimes made for articles of clothing in the bill of outfit.* Some astounding facts might be mentioned illustrating this point; but we let them pass, hoping they will never be reënacted again. Besides, instances have been known, in which there was a sad discrepancy between the seaman's bill of clothing, and the number of articles actually found in his chest, when he first examined it, after the ship was got under weigh, and bound out to sea.

It is when whalemen are beyond our coast, and around the Horn, and their outfits have been put to some service, they find that the winds, storms, and exposures have made sad havoc of their supposed sound and reliable chest of clothing. The fact that they are not present, but absent, and will be for months and years, and therefore unable to speak for themselves, face to face, to those by whom they have been *sold*, poorly and wretchedly justifies frauds, which may not be heard from for months, and perhaps for years. Time, however, stereotypes, instead of obliterating, a wrong. And still further, extravagant and unreasonable inducements and promises held out to influence thoughtless youth to engage in the business of whaling, are connected with the evils which have grown out of the system of outfitting, and of which whalemen and others have justly complained. These evils, however, if they now exist, could be measurably removed, if agents of the respective ships would carefully examine the bill of clothing which each seaman brings from the outfitter, article by article, contained in his chest; or if captains and officers should take this thing in hand after the sailing of their ships from port, and thus ascertain from personal inspection whether their crews have been justly dealt with as to the quality and number of articles in their bill of outfits.

Were this course thoroughly pursued, it would put at once, we are free to assert, a wholesome check upon any further attempts to defraud the ignorant and unsuspecting. We may go even yet further, and say, that it is clearly the *duty* of agents and officers of ships to look after the interests of inexperienced seamen who sail in their employ, and under their command; and if they did as suggested, it would doubtless greatly conduce to the contentment of seamen on shipboard, and likewise promote mutual good will and understanding, in regard to the purposes of the voyage.

It is hoped, however, there is less disposition now than formerly to defraud the unsuspecting, either in the quality or number of articles included in their outfit, or to deceive the ignorant by presenting to their minds unreasonable and extravagant promises, which would never be realized; and that honesty, which is always the best and safest policy at all times, and under all circumstances, will henceforth be more obviously seen in this branch of business connected with the whale fishery.

The following is an inventory of the principal articles of clothing, and a few incidentals, included in a young whaleman's outfit for a voyage, of two, three, or four years. It may serve as a sort of directory, and thus be of considerable advantage to those who would prefer a good, substantial outfit, compared with one which may have a name simply, but deficient in a great measure in real worth and service.

1 Monkey Jacket,	4 Undershirts,
1 Reefing do.,	6 Pairs Thin Pants,
1 Oil Suit,	6 Thin Shirts,
4 Pairs Thick Pants,	4 Pairs Thick Drawers,
6 Thick Shirts,	1 Guernsey Frock,

1 Thin Frock,	1 Tin Pot,
6 Pairs Shoes, good,	1 Spoon,
6 " Stockings,	1 Tin Pan,
6 Jackknives.	1 Bed, (mattress,)

Quilt, Blanket, Pillow, &c., &c.; Razor, Strop, Soap, Needles, Thread, Brush, &c.

The chief purpose the writer had in view in bringing together these facts connected with the history and details of whaling, was not only to interest the general reader, but that young men from the country, and elsewhere, who are desirous of engaging in this branch of employment, may know somewhat of its character and pecuniary importance in a commercial point of view.

Men of the first business talents are enlisted in this enterprise at home; and a more hardy, thorough-going, energetic, and generous class of men, as captains and officers of ships, do not traverse the ocean. That there are exceptions to this general rule, none will deny. Capriciousness, tyranny, crossness, and inhumanity are exhibited by some upon the sea as well as upon the land. He who cannot govern himself is ill prepared to be the leader of others.

It may be said that whalemen are at home on the ocean. During the first fifteen or twenty years of their service, they scarcely remain at home with their families and friends as many months. It is no mean and unworthy profession, but one highly honorable and creditable for any aspirant.

Nor is the responsible position of a captain, or officer, attained at once. Promotion comes not from the cabin windows, but in a direct line from the forecastle. There must be a regular apprenticeship gone through with, before one can expect to succeed in the hazardous undertaking of capturing the monsters of the deep. It is a *trade*, and in this regard it is far different from the merchant service. In addition to good seamanship, — and, by the way, whalemen are acknowledged to be among the best navigators in the world, — it is *whale killing*, an aptitude for this particular kind of work, that gives promise of attainment and success in the profession.

This business, then, we say, holds out many reasonable inducements to a young man desirous of engaging in it. With a good common school education, energetic, faithful to himself and his employers, temperate, and withal having a purpose to be something and do something in the world, there are but few paths to honorable respect, character, influence, and pecuniary competence, more inviting than this.

There are trials, and peculiar ones too, in the whaling service; and in what branch of industry are there not? But making all the allowances for long absences from home, which, without doubt, are the greatest deprivations of all, yet there are other considerations, which, it is believed, counterbalance these disadvantages.

There are dangers also connected with whaling; aside from the storms and sufferings which whalemen experience in navigating those remote northern seas and oceans, the greatest exposure to life is doubtless in the work of whaling. Yet, taking into the account the number of vessels and seamen engaged in this business, the distant places visited by them, and the character of their employment, and we venture the assertion, that there is no department of commercial enterprise, whether coastwise or foreign, that can present a list more free from disaster, loss of life, or bad health among seamen, than the whaling fleet.

While varied success attends the labors and deprivations of whalemen, yet, on the whole, we must conclude that the enterprise is as profitable, and furnishes as strong inducements for the investment of capital, as almost any other.

There have been partial and individual reverses in the whaling business, it is true, and unforeseen contingencies will ever happen; yet this fact is most obvious and plain to be seen, that from the whole history of whaling in this country, those seaport places in which the business has been perseveringly carried on, will advantageously compare with inland manufacturing and farming communities, in enterprise, wealth, educational appliances, and in all the comforts, and even the luxuries, of life.

CHAPTER VII.

Manufacture of Oil.

CRUDE oil, or oil in its natural state, is that which is obtained from the blubber of the whale in the process of "trying out" on shipboard. The oil, then, which is taken from whale ships and carried to the oil manufactory, is said to be in its *crude* state. We will speak first of the manufacture of *crude sperm oil.**

The first step in the process of manufacture, is to take the oil in its crude state, and put it into large kettles, or boilers, and subject it to a heat of one hundred and eighty to two hundred degrees, and then all the water which happened to become mixed with the oil, either on shipboard or since, will evaporate.

WINTER STRAINED SPERM OIL. In the fall, or autumn, the oil is boiled for the purpose of granulation during the approaching cold weather. The oil thus passes from a purely liquid into a solid state, or one in which it is in grains, or masses.

When the temperature of the atmosphere rises, or the weather slackens during the winter, the oil which has been frozen, but is now somewhat softened, is shovelled out of the casks and

* The author is indebted to Charles J. Barney, Esq., foreman of Dr. Daniel Fisher's oil factory, in Edgartown, one of the largest, if not the largest in the country, for the principal facts respecting the manufacture of oil.

BOILING OUT.

put into strong bags that will hold half a bushel or more, in order to be pressed. The oil which is now obtained from this first pressing is called *winter strained sperm oil.*

Spring Sperm Oil. What remains in the bags after the first pressing, is again heated by being put into boilers, after which it is baled into casks again, and upon cooling, it becomes more compact and solid than it was before.

During the month of April, when the temperature is about fifty degrees, the oil becomes softened ; it is then put into bags, and goes through a second process of pressing similar to the first. The oil from this pressing is called *spring strained sperm oil.*

Tight Pressed Oil. That which is left in the bags after the second pressing, is again melted, and put into tin pans or tubs which will hold about forty pounds each. When this liquid is thoroughly cooled, as each pressing makes what is left harder, in consequence of extracting the oil, the cakes taken from the tubs are then carried into a room heated to about ninety degrees; and as they begin to yield to the influence of this high temperature, or the remaining oil begins to soften the cakes, they are taken and shaved into very fine pieces, or ground up as in some instances, deposited in bags as hitherto, and put into the hydraulic press.

The room being at the temperature indicated above, and the bags subjected to a powerful pressure of three hundred tons or more, all the oil is extracted from them, and what is left is perfectly dry, free from any oily matter, and brittle. The oil thus obtained by this last pressing is called *tight pressed*, or *summer oil.*

Spermaceti. What remains after the several pressings, and the removal of all the oil, is called *stearine*, or *spermaceti.*

Spermaceti is not confined to the head matter of the whale, as some suppose, nor does the head matter have any thing to do with the brains of the whale, as others have falsely conjectured; but spermaceti is found in the entire oil from the sperm whale.

It should be observed, however, that the spermaceti from the body oil of the whale makes harder candles than the spermaceti from the head matter; but the head oil or matter gives a greater proportion of spermaceti, and is more valuable than that from the body oil. Besides, the spermaceti from the head oil is quite different from that of the body oil; the former presents fine, bright, transparent scales, like small particles of isinglass, while the latter is more compact, something like dough. In cooling, one exhibits a sparry, crystalline structure, the other that of clay.

Head oil or matter is usually manufactured with the body oil of the whale, and mixed in proportion to one third of the former to two thirds of the latter.

Spermaceti Candles. That which remains in the bags after the hydraulic pressure is both dry and brittle. The oil, it is supposed, is wholly extracted, and nothing now remains but the spermaceti. Its color, however, is not white, but interspersed with grayish streaks, bordering on the yellow.

The spermaceti is put into large boilers adapted for the purpose, and heated to the temperature of two hundred and ten degrees. It is refined and cleared of all foreign ingredients by the application of alkali. Afterwards water is added, which, with a temperature of two hundred and forty degrees, throws off the alkali in the form of vapor. The liquid which remains is as pure and clear as the crystal water, and ready to be made into the finest spermaceti candles.

Right Whale Oil. The manufacturing of this variety of oil is of recent date, (within twenty-five years.) At first, in preparing it for sale, it was taken in its crude state and "reeked off," that is, simply pumped out of the casks, and leaving the sediment behind. This kind of oil then was as cheap as milk is now.

Bleaching Oil. Crude oil is bleached in the first place by putting it into large kettles, applying alkali to it in proportion of one quart of alkali to one barrel of oil, and then heating it to a temperature of one hundred and ten degrees. This process destroys the watery acid in the oil.

WINTER STRAINED WHALE OIL. Whale oil, after bleaching, is made into winter oil by exposing it in casks to cold weather, or by artificial freezing in the summer. When frozen it is granulated, or separated into grains, or masses, like sperm oil. At a temperature when the oil begins to exhibit liquid particles, it is taken from the casks, and put into double cotton strainers. The oil which comes from this straining is called *winter strained whale oil.*

The following facts respecting winter strained whale oil may not be wholly destitute of interest. It is found that it will endure a greater degree of cold at the same temperature than winter pressed sperm oil; it will burn longer, and its specific gravity is heavier; but it will not give so brilliant a light as sperm oil.

SPRING WHALE OIL. What remains after straining the first time, goes through the process of heating, cooling, and pressing, similar to spring sperm oil; and thus is obtained the *spring whale oil.*

That which is left after straining and pressing is called *whale foots.*

The following are a few of the uses to which whale foots are applied. In making an inferior kind of candles, in making some kinds of bar soap, and likewise used on railways and in ship yards.

The *adamantine candles* are made of spermaceti mixed with wax, in proportion of one ounce of wax to a pound of spermaceti, and subjected to powerful steam pressure. They are not only much harder than spermaceti candles, and variously colored, but they command a higher price in the market. There is a manufactory of this description in Philadelphia.

Oil soap is made from the deposit of alkali, in the process of bleaching. If, after pressing and bleaching, the oils should retain too dark a color, they are then bleached again. Some varieties of oil are darker than others, which requires additional labor in this respect.

There is another method, and usually the ordinary one, by which oils are clarified and prepared for the market. It is termed *panning*.

For this purpose, after it has been bleached, strained, or pressed, and it does not assume the right color or shade, it is pumped into large, leaded, superficial vats, or pans, located in a building near by, whose side roof is wholly of glass, like a glass house, and so arranged that both air and sun can act upon large bodies of oil in different stages of whitening.

This process not only whitens the oil, but whatever particles or thickness there may have been in the oil, not discernible before, is now all removed and deposited on the bottom of the pan. The oil taken from these pans is put into barrels or casks, and is ready for the market.

Government Test of Sperm Oil. The lighthouses upon our seaboard, and also upon the lakes, are furnished with the best quality of sperm oil.

Sperm oil has a standard weight established or recognized by the government, and according as varieties fall short or go beyond this measure, or standard, indicated by a nicely adjusted oilometer, its true weight and value are ascertained.

Winter strained sperm oil is heavier, and burns away faster than spring strained sperm oil, for the simple reason that the winter oil is freer from spermaceti than the spring strained oil.

Sperm oil is tested by authority of government, when contracts are made with the manufacturer to furnish oil for lighthouses, in the following manner: A common tin lamp of a cylindrical form is taken, and fitted with a wick which reaches to the bottom of the lamp. It is then filled with oil, and kept burning until all the oil in the lamp is burned up, and the wick so dry that not a drop of oil will fall from it.

The number of hours which a given quantity of oil will burn, is another consideration included in determining the relative quality of oil.

MIXING OR ADULTERATING OILS. Sperm oil is the purest and best of all varieties of whale oil, and brings the highest price in the market. Sperm oil is frequently mixed in greater or less proportions with right whale oil, which is an inferior kind of oil, and labelled *sperm oil.* This is a fraud, and it is practised more extensively than people are aware of.

The fraud may be detected if either the right whale oil or sperm oil happens to differ in its shade or color, the one from the other. Sperm oil is lighter than right whale oil. If sperm oil is carefully put into a glass containing right whale oil, the former will not displace the latter, but remain separate; and the line of separation between the two kinds of oil, providing the color is somewhat different, may be easily detected. The smell and taste of oil, likewise, determine whether it be *bogus* sperm oil or not.

But the surest and most certain test of all others, as to the quality of oils, is by the oilometer,* (elaiometer.) This instrument is not only authorized by government, and employed for the purpose of securing genuine sperm oil, but it is used in all oil manufactories to determine both in buying and selling the different varieties of oils.

It should be observed, however, that the mixing or adulteration of oils is never practised by the oil manufacturer. It would be fatal to his business if he should do it. There is no such article in an oil factory for sale as mixed or adulterated oils. It is either sperm or otherwise. The mixing takes place after it passes into the hands of the second, third, and it may be the fourth purchaser, or retailer. It is believed that but little genuine sperm oil reaches far back in the country, except, perhaps, for the purpose of lubrication in machinery; and even then, much of that, if tested by the proper measure, would probably be found badly mixed with an inferior quality of oil.

Since preparing the present chapter on the manufacture of

* Harris.

oil, the following just remarks came under our observation, selected from the *New Bedford Mercury*, January 16, 1857, and are worthy of the attention of all.

"We embrace the present opportunity to offer some remarks and suggestions in the matter of the adulteration of sperm oil, which has been carried on to such an extent as to form one of the causes, we may safely say, for the decline of the article in price. Like every other commodity, it is liable to be counterfeited; and we know, after it has passed out of the hands of importers and manufacturers, adulteration has been practised to a wide extent. This is an evil beyond the control of our merchants, however much they may deprecate its influence upon the trade. To say that a system of adulteration is practised here, is a charge which cannot be substantiated by facts; and to suppose that the manufacturer would knowingly injure or damage his own business, is too absurd to require refutation. . . . We believe it would be a wise and judicious policy to establish agencies in different sections of the country, with agents of known honesty and integrity, for the wholesale and retail of the article; and that each barrel and cask bear the name of the manufacturer as a guaranty of its purity.

"Consumers, especially those who buy for machinery purposes, would then know where to make their purchases, and have the assurance of 'value received.'"

www.ingramcontent.com/pod-product-compliance
Lightning Source LLC
Chambersburg PA
CBHW021946220326
41599CB00012BA/1201
9783337253189